Local water partnership

Local water partnership

Universal model of water resources management

By

Jarosław Gryz
Sławomir Gromadzki

BRILL | WAGENINGEN ACADEMIC

Cover illustration: Photo by Jarosław Gryz

The Library of Congress Cataloging-in-Publication Data is available online at https://catalog.loc.gov
LC record available at https://lccn.loc.gov/2024940148

Typeface for the Latin, Greek, and Cyrillic scripts: "Brill". See and download: brill.com/brill-typeface.

ISBN 978-90-04-69605-1 (hardback)
ISBN 978-90-04-69906-9 (e-book)
DOI 10.3920/9789004699069

Contents

Summary 161

Figures, maps and tables

Figures

Maps

Tables

Introduction

Water resource, any of the entire range of natural waters that occur on Earth regardless of their state (i.e., vapour, liquid, or solid) and that are of potential use to humans. Of these, the resources most available for use are the waters of the oceans, rivers, and lakes; other available water resources include groundwater and deep subsurface waters and glaciers and permanent snowfields.[1]

∴

Human existence on Earth is implied by its environment, natural and social. This obvious relationship is expressed differently than before, in past millennia. In the new geological era, the Anthropocene, environmental changes include the impact of humans enabling or not enabling the survival of plants, animals, humans, civilisations. Given the magnitude of the challenges accompanying this phenomenon, the transformation of human activities that reduce the negative changes occurring in the biosphere takes precedence. Above all, those that involve environmental protection, conservation and restoration, associated with agriculture and the economy; and furthermore, human existence itself on Earth.

Conservation of water resources in the European Union is part of the global climate puzzle. Indeed, the puzzle depends both on what is done 'here and now' and the impact that the organisation's actions and those of its member states will have on the decades and centuries to come. When taking into account the projection of planetary-scale change, the 'European climate puzzle' may seem insignificant, but it forms a snapshot of the whole. More than that, they condition the survival and development of societies living between the Oceans: The Atlantic and Arctic, and the Mediterranean Sea. Their impact is significant not only on the continent situated here, but also the other areas around it. The challenges for the European Union here are climate change, sustainable development, social cohesion, security in the broadest sense (ecosystem, digital, technological) within the framework of the goals (sustainable development goals) that the United Nations has embraced.

Humanity's civilizational challenges in the 21st century is related to reducing the adverse impact of humans on planet Earth. The policy that defines the approach to these challenges is the European Union's Green Deal. It has its own ideological, social, and economic rationale. Its aim is to make the European continent the first place on the planet where the carbon footprint from industry, services, agriculture, transport, the lifestyles of the community's citizens will be negated or completely

---

[1]   Britannica, accessed 20.10.2022, https://www.britannica.com/science/water-resource.

eliminated. The projected timescale for the transformation is 30 years, from 2020 to 2050, thanks to the compliance of EU policies with the Paris Agreement, accelerating the transition to renewable energy sources and increasing energy efficiency, reducing dependence on external sources, diversifying supply and investing in solutions for future mobility, improving air and water quality, promoting sustainable agriculture, implementing the European pillar of social rights at EU and Member State level, strengthening climate action. This is done with reference to United Nations *Transforming our world: the 2030 Agenda for Sustainable Development* approved in year 2015.

The basis of the European Union's Green Deal, embedded in the framework of green capitalism, is the premise that Europeans' hitherto high standard of living will be improved by responding to vital individual, group, societal and ecological demands. In addition, it will allow capital markets to continue operating as they have done to date. In this context, the projection of the European Union's policies, directives and programmes includes, and is expected to reinforce, several parallel processes defining 'green capitalism', i.e., ecological ways of producing, consuming, and recycling the residues of goods used. Consequently, this will lead to several key changes in the economy of this area of the planet. Firstly, there will be a decoupling of the increasing EU Member States' GDP from environmentally destructive, increasing energy consumption. Secondly, there will be an EU-wide roll-out of environmentally neutral energy generation and usage technologies. Thirdly, the complete elimination of energy consumption impacting on the environment will be achieved. Fourthly, waste treatment will be almost entirely eco-friendly. The European Green Deal achieves all these objectives. Its key elements are:
- a circular economy, i.e., 'green' jobs provided by 'green' consumption and 'green' recycling of residues in a 'green' cycle of energy extraction and usage (organic food, energy-saving technologies, goods made from recycled waste are and will be the result),
- conservation of the environment and biodiversity in Europe,
- the farmer's 'field to table' consumer food consumption strategy.

As a result of this approach, certain types of policies are to be implemented:
- increasing the competitiveness of green European industry,
- ensuring the transformation of regions affected by the transformation of the previous economic order, their shift away from the use of fossil fuels, non-renewable energy,
- changing the skills of workers affected by the transformation of the previous economic order,
- the creation of 'green', environmentally aware communities in European Union countries, aware of their individual and group roles in the processes of consumption and production of goods.[2]

---

2 The European Green Deal, communication from the Commission to the European Parliament, the European Council, the Council, the European Economic and Social Committee and the Committee of the Regions, Brussels, 11.12.2019 COM(2019) 640 final.

In the European Union's approach, one of the most essential elements of the environmental puzzle is the conservation of water resources, and their retention as an instrument for shaping the European Green Deal. However, to date, the practice of the European Commission and the European Council has not incorporated either a preferred model or even a vision of it. This can be attributed to the sharing of powers in this area between different institutions. In addition, the fragmented nature of activities related to this conducted in individual countries. This factor influences the framing of EU policies, directives, strategies and programmes.[3] It also conditions the development of the concept of a universal model for water conservation, and its retention in terms of agricultural, non-agricultural activities, and their impact on climate issues. The existing European Union Water Framework Directive and its accompanying documents have not introduced regulations in this area that would comprehensively address the implementation of solutions linked to these issues. More importantly, the potential that exists in this area has not yet been properly identified and verified in terms of its usefulness for the New Green Deal and the action of European Union institutions in this context. This is especially relevant as it covers geographically and climatically diverse regions that are part of it. If we relate this to the European Union, we should expect a reduction in annual rainfall; its sudden, abrupt form; prolonged droughts, and other natural disasters.

The EU Climate Change Adaptation Strategy 2013 set out a framework and mechanisms to better prepare the EU for the current and future impacts of climate change. It proposed to achieve this by supporting and stimulating EU Member State action on climate change mitigation, by laying the groundwork for better informed decision-making on resilience in the coming years, and by making key economic and political sectors more resistant to the effects of climate change. In doing so, it has placed emphasis on the following activities:
- encouraging all Member States to adopt comprehensive adaptation strategies.
- Providing LIFE funding to support capacity building and accelerate the pace of climate adaptation activities in Europe.
- mainstreaming adaptation within the Covenant of Mayors in cities, particularly through voluntary commitments to adopt local adaptation strategies and awareness-raising activities.
- Addressing knowledge gaps related to:
  - information on the consequences, costs and benefits linked to adaptation.
  - local and regional risk analyses and assessments.
  - frameworks, models and decision-support tools to assess the effectiveness of different adaptation measures.
  - means of monitoring and evaluating adaptation measures to date.
- Further develop Climate-ADAPT as a 'one-stop shop' for climate change adaptation information in Europe.

---

3 Decision (EU) 2022/591 of the European Parliament and of the Council of 6 April 2022 on a general union environment action programme to 2030, Official journal of the European Union, 12.4.2022.

- Facilitating the resilience of the Common Agricultural Policy (CAP), Cohesion Policy and Common Fisheries Policy to climate change.
- Providing a more resilient infrastructure.
- promoting insurance and other financial products to assure climate-resilient investments and commercial decisions.

They form a framework that can be completed with content through the adaptation of a water management system, of which water retention and conservation at the local level will become an integral part.

The document Clean Planet for All … identifies eight scenarios to reverse the identified trend. **None of them refer to a model for water conservation, water retention, applicable across the whole of the European Union.** They all centre on reducing greenhouse gas emissions either directly or indirectly. An update on the European Union's initiatives for comprehensive climate protection follows:

- The financial sector in the service of the climate.
- The European external investment plan – opportunities for Africa and the EU neighborhood region.
- Support to cities for investment in urban areas.
- Clean Energy for Islands initiative.
- Structural support for regions with high coal production and high carbon emissions.
- European initiative 'Young people for the climate'.
- Smart building investment financing instrument.
- A compilation of EU regulations on the energy performance of public buildings.
- Investment in clean industrial technologies.
- Clean, competitive and network-based mobility.

None of the instruments indicated per se addresses the model of water conservation, water retention and agricultural use, industrial activity or flood and drought prevention. A group of intermediate measures reducing CO_2 emissions as well as reducing the carbon footprint are indicated. It should therefore be noted that the European Union does not have specific approaches on how to combine water conservation and retention. As things stand, there are too many European Commission projects, and this results in too many widely fragmented policies in this area. This primarily arises through the specific profile of climate investment in the environment of the Member States, i.e., the elimination of CO_2 from the atmosphere through its absorption, the purification of water, and the enhancement of biodiversity in the environment.

Three conclusions emerge from an assessment of the character of the policies implemented by the European Commission. Firstly, the wide variety of countries in the Union with different climatic conditions, e.g., maritime countries vis a vis landlocked country with no access to the sea or the ocean, lowland countries vis some vis those in the mountains, results in the dispersion of resources. This should be changed to match the specific characteristics of these needs identified in the water conservation model. Secondly, the security of the EU and its Member States as a factor integrating policies and strategies with an emphasis on the development of a green economy based on the protection of water resources creates hitherto unidentified development opportunities. It allows for the involvement of individual countries as well as the entire organisation in a comprehensive, Europe-wide effort. Third, that the current weaknesses of the current organisation of EU action in terms of the relationship between crisis management systems and water resource protection systems including flood measures and drought mitigation are due to:

- separate operational systems in individual EU countries,
- activities pursued under different regulations,
- the dispersion of competences in EU bodies as well as member state administrations,
- difficult coordination of activities, policies and strategies at different levels of EU administration, their respective competences and ability to create synergies,
- disorganised and inconsistent management of the water resources of EU countries,
- the irrationality and wastefulness of actions adopted and implemented by public institutions,
- the lack of a systemic effect encompassing the mission, vision and related objectives of Member States in the field of water conservation within the European Union.

The considerations undertaken in this paper define a new potential direction for the European Union and, at the same time, a paradigm for thinking about environmental[4] and water security.[5] Considering the ambitious goals of the organisation, the solutions indicated here can serve as a benchmark for the organisation's policies, strategies, directives, and programmes. All the more so as the EU has specialised instruments at its disposal, such as the European Fund for Strategic Investment, cohesion policy funds and the European Agricultural Fund for Rural Development (EAFRD), which can easily be adapted to the needs of water resource conservation throughout the EU, albeit with different effects in individual regions. The study addresses the realisation of the expectations of society, the economy,

---

4   Ecological security – preventing the destruction of land, air, and water resources, various types of pollution, habitat loss, deforestation, desertification, loss of biodiversity, threats to individual species, ozone depletion, destabilisation and degradation of terrestrial and marine ecosystems. A term derived from environmental safety understood as environmental security. Lorraine Elliott, "Human security/environmental security," Contemporary politics, 21:1 (2015): 11–24, https://doi.org/10.1080/13569775.2014.993905.

5   Hydrosecurity – rational and efficient use of water resources, limiting water losses ensuring environmental, food and economic security while taking into account coherent counteraction of flood and drought threats within the crisis management system. S, Gromadzki, R. Panfil (term defined during the scientific conference Contemporary determinants of security and logistics in the local, regional and global environment), PUZIM Ciechanów, Mława 10.05.2022.

including agriculture, public institutions and states, the policies and strategies of the European Union and other social actors and international organisations.

The book adopts a theoretical approach based on governance in European Union institutions at a time when their agenda is changing.[6] A distinction is made here between approaches involving time, space, social phenomena, and relating them to the European Union. Existing patterns of cooperation and persuasion indicated the range of alternatives available for the implementation of the water retention model in the form of: Community institutions, Member State government institutions, civil society institutions, as well as farmers/entrepreneurs. In doing so, a heuristic method was adopted based on the opinions expressed in works relating to the various policy dimensions, mechanisms and processes of change in relation to the natural environment and human existence. Particularly in terms of impact in a bottom-up approach with European Union social actors. Thus, the scientific interpretation applied in this study is intended to enable the practical application of knowledge linked to the concept of a European / Community universal water retention and conservation model throughout the whole European Union. In doing so, it applies to a responsible citizen, an entrepreneur who shapes the environment with their behaviour and fosters trends in its protection.

The work is based on an interpretation of a body of knowledge that compiles many of the interdependent issues determining the relationship between climate change, environmental protection, economic development and the projection of change in these areas resulting from the European Union's water conservation model. The starting point is climate change determined by human activity and its impact on water retention, water conservation, drought prevention[7] and the adaptation of new technologies to assist agriculture in this area.[8] This relates to landscape transformation, loss of soil productivity, maintenance of functioning ecosystems, carbon emissions, biodiversity, desertification and contamination of water resources, and the capacity to develop sustainable agriculture.[9] A report by the UN Intergovernmental Panel on Climate Change (IPCC) clearly shows the link between agricultural practices and the loss and degradation of natural resources.[10]

---

6   Carmen Maganda, "Water security debates in 'safe' water security frameworks: moving beyond the limits of scarcity," Globalizations, 13:6 (2016): 683–701.

7   Thomas Berger, Regina Birner, Nancy Mccarthy, Jose Díaz, Heidi Wittmer, "Capturing the complexity of water uses and water users within a multi-agent framework," Water resources management 21 (2007): 129–148.

8   David Christian Rose, Jason Chilvers, "Agriculture 4.0: broadening responsible innovation in an era of smart farming," Frontiers in sustainable food systems 2 (2018): 1–7, https://doi.org/10.3389/fsufs.2018.00087.

9   World Bank. 2021, "Water in agriculture," World Bank, accessed 20.03.2022, https://www.worldbank.org/en/topic/water-in-agriculture#1.

10   Climate change and land, an IPCC special report on climate change, desertification, land degradation, sustainable land management, food security, and greenhouse gas fluxes in terrestrial ecosystems, accessed 11.11.2020, https://www.ipcc.ch/srccl/.

As a consequence of changes brought about by the current approach, alterations are needed in all types of land management[11] from forestry to fisheries to livestock, a transition to sustainable agricultural practices to restore landscape integrity and productive capacity.[12] In the interpretation adopted in the study combining a change of approach with new technologies, is the aspect that led our thinking.[13]

The underlying consideration is climate change and its implications for societies in relation to sustainable agriculture.[14] Agriculture, requiring the use of new technologies for its conservation and regeneration.[15] Particularly those related to data acquisition, and data processing for the promotion of an appropriate approach to cultivation at the scale of the individual farm[16] as well as the agricultural sector as a whole.[17] This allows insight into how an individual farm, and its sector, functions and how it impacts, among other things, on climate change and,[18] furthermore, on reducing greenhouse gas emissions.[19] Therefore, when identifying solutions for the development of agriculture in this study, the emphasis is on those related to its digitisation as the backbone of the technological revolution[20] and moreover, the

11   Woldegebrial, Zeweld, Guido Van Huylenbroeck, Girmay, Tesfay, and Stijn Speelma, Smallholder farmers' behavioural intentions towards sustainable agricultural practices, Journal of environmental management 187 (2017): 71–81, DOI: 10.1016/j.jenvman.2016.11.014.

12   Shira Bukchin, Dorit Kerret, "The role of self-control, hope and information in technology adoption by smallholder farmers – a moderation model." Journal of rural studies 74(4) (2020): 160–168, DOI: 10.1016/j.jrurstud.2020.01.009.

13   Annamaria Castrignano, Gabriele Buttafuoco, Raj Khosla, Abdul Mouazen, Dimitrios Moshou, Olivier Naud, eds., Agricultural Internet of Things and decision support for precision smart farming, (Academic Press, 2020).

14   Mohammed J. Taherzadeh, Bioengineering to tackle environmental challenges, climate changes and resource recovery, Bioengineered, 10:1 (2019): 698–699.

15   Long Le Hoang Nguyen, Arlence Halibas, Trung Quang Nguyen, Determinants of precision agriculture technology adoption in developing countries: a review, Journal of crop improvement (2022): 11–13.

16   Laurens Klerkx, Emma Jakku, Pierre Labarthe, A review of social science on digital agriculture, smart farming and agriculture 4.0: new contributions and a future research agenda, NJAS: Wageningen, journal of life sciences 90–91:1 (2019): 3–6.

17   Alm E. Colliander, Lind F. Stone, V. Sundström, O. Wilms & M. Smits, Digitizing the Netherlands: how the Netherlands can drive and benefit from an accelerated digitized economy in Europe. Boston Consulting Group, 2016.

18   Margaret Ayre, Vivienne Mc Collum, Warwic Waters, Peter Samson, Anthony Curro, Ruth Nettle, Jana-Axinja Paschen, Barbara King, Nicole Reichelt, "Supporting and practising digital innovation with advisers in smart farming," Njas – Wageningen journal life sciences, Vol. 90–91 (2019): 1–11.

19   Athanasios Balafoutis, Bert Beck, Sypros Fountas, Jurgen Vangeyte, Tamme Van Der Wal, Iria Soto, Manuel Gómez-Barbero, Andrew Barnes, Vera Eory, "Precision agriculture technologies positively contributing to ghg emissions mitigation, farm productivity and economics sustainability," Sustainability 9(8), 1339 (2017): 1–21, doi.org/10.3390/su9081339.

20   Jonathan Steinke, Jacob van Etten, Anna Müller, Beta Ortiz-Crespo, Jeske van de Gavel, Silvia Silvestri, Jan Priebe, "Tapping the full potential of the digital revolution for agricultural extension:

adaptation of innovations.[21] In terms of the changes presented in the study, the implementation of digitisation and the development of agriculture, with its associated challenges, became the foundation.[22,23] The interdisciplinary of the interpretation contained in the text of the study[24] is what is indicative of the priorities for thinking and acting.[25] Especially the stakeholders themselves, i.e. the farmers,[26] their approach to technological changes in agricultural production and the protection of natural resources.[27] The attitude of the aforementioned groups was assumed to be of particular importance for the implementation of IT-based Agriculture 4.0.[28] The issue at stake here is not only the researchers or farmers themselves, but the entire social environment associated with their activities.[29] In doing so, it relates this activity to specific conditions which, by virtue of expertise, are embedded in the realities of agricultural transformation.[30] In doing so, the formula of change included in the study relates to multiple levels of information, social, and political

an emerging innovation agenda," International journal of agricultural sustainability 19:5–6, March (2020): 549–565, DOI: 10.1080/14735903.2020.1738754.

21   Evangelos D. Lioutas, Chrysanthi Charatsari, "Innovating digitally: the new texture of practices in agriculture 4.0," Sociologia ruralis 62:2 (2022): 250–278, DOI: 10.1111/soru.12356.

22   Emma Jakku Simon Fielke, Aysha Fleming, Cara Stitzlein, "Reflecting on opportunities and challenges regarding implementation of responsible digital agri-technology innovation," Sociologia ruralis 62:2, (2022): 363–388, DOI: 10.1111/soru.12366.

23   Áline Regan, "Exploring the readiness of publicly funded researchers to practice responsible research and innovation in digital agriculture," Journal of responsible innovation, 8:1, (2021): 28–47, https://doi.org/10.1080/23299460.2021.1904755.

24   Andrea Schikowitz, "Creating relevant knowledge in transdisciplinary research projects – coping with inherent tensions," Journal of responsible innovation 7 (2) (2020): 217–234, https://doi.org/10.1080/23299460.2019.1653154.

25   Mark Shepherd, James A. Turner, Bruce Small, David Wheeler, "Priorities for science to overcome hurdles thwarting the full promise of the 'digital agriculture' revolution," Journal of the science of food and agriculture 100 (14) (2018): 5083–5092, DOI: 10.1002/jsfa.9346.

26   Kelly Bronson, "Smart farming: including rights holders for responsible agricultural innovation," Technology innovation management review 8 (2) (2018): 7–14, http://doi.org/10.22215/timreview/1135.

27   Kelly Bronson, "Looking through a responsible innovation lens at uneven engagements with digital farming," NJAS–Wageningen journal of life sciences 90–91 (2019), 3–5, https://doi.org/10.1016/j.njas.2019.03.001.

28   David Christian Rose, Jason Chilvers, "Agriculture 4.0: broadening responsible innovation in an era of smart farming," Frontiers in sustainable food systems 2 (87) (2018): 1–5, https://doi.org/10.3389/fsufs.2018.00097.

29   Margaret Ayre, Vivienne Mc Collum, Warwick Waters, Peter Samson, Anthony Curro, Ruth Nettle, Jana-Axinja Paschen, Barbara King, Nicole Reichelt, "Supporting and practising digital innovation with advisers in smart farming," NJAS: Wageningen journal of life sciences 90–91:1 (2019): 1–12, 10.1016/j.njas.2019.05.001.

30   Krzysztof Janc, Konrad Czapiewski, Marcin Wójcik, "In the starting blocks for smart agriculture: the internet as a source of knowledge in transitional agriculture," NJAS: Wageningen journal of life sciences, 90–91:1 (2019): 1–12, https://doi.org/10.1016/j.njas.2019.100309.

systems.[31] In the identified processes, data was treated as an integrating factor, from the individual farmer to the implications of their activities for all other social, economic, and political stakeholders.[32] In doing so, special attention was paid to public-private partnerships.[33] It was recognised that only a comprehensive approach by farmers, forest owners, public institutions, and the public would provide a guarantee of an appropriate approach to climate change-related agricultural transformation.[34]

The model for the conservation of water resources within the European Union presented in this study is linked to water security, which includes further areas not defined in this way. This is because it relates to measures that determine the capacity to safeguard water, to counteract floods, droughts, and various hazards that occur over time and space.[35] Here, a range of interdependent issues play a special role. Foremost amongst these is decarbonisation. Innovative agriculture plays a special role in this and can occur as a reinforcing element in modelling the European Union's actions, policies and strategies as part of the Green Deal. This element is important in determining the costs, direct and indirect, associated with CO_2 implications of agricultural production, industry as well as the social development of Member States. The conservation of water resources creates the right conditions not only to reduce CO_2, but also to strengthen the European economy, including agriculture, in the farmer-consumer relationship, in short food supply chains. In addition, strengthening research work related to water conservation technologies, for the wider wellbeing of societies.

The subject of this study is the implementation of the New Green Deal, which implies the European Union's actions related to the realisation of its objectives for the development of societies' well-being ensuring a secure existence, which is determined by access to water resources, their use and the maintenance of their quality. In doing so, it identified proposals for organisational and normative changes, and the competences and responsibilities of public, private and social stakeholders

---

31   Simon Fielke, Robert Garrard, Emma Jakku, Alysha Fleming, Leanne Gaye Wiseman, Bruce M. Taylor, "Conceptualising the DAIS: implications of the 'Digitalisation of Agricultural Innovation Systems' on technology and policy at multiple levels," NJAS: Wageningen journal of life sciences, 90–91(3) (2019): 1–3, DOI: 10.1016/j.njas.2019.04.002.

32   Michael Carolan, Digitization as politics: smart farming through the lens of weak and strong data. *Journal of rural studies* 91 (2022): 208–216, https://doi.org/10.1016/j.jrurstud.2020.10.040.

33   Dennis Pauschinger, Francisco R. Klauser, "The introduction of digital technologies into agriculture: space, materiality and the public–private interacting forms of authority and expertise," *Journal of rural studies* 91 (2022): 217–226, https://doi.org/10.1016/j.jrurstud.2021.06.015.

34   Anja. Bauer, Alexander Bogner, Daniela Fuchs, "Rethinking societal engagement under the heading of responsible research and innovation: (novel) requirements and challenges," Journal of responsible innovation 8:3 (2021): 342–363, https://doi.org/10.1080/23299460.2021.1909812.

35   Hector Ibarra, Jerry Skees, "Innovation in risk transfer for natural hazards impacting agriculture," Environmental Hazards 7:1 (2007): 62–69, https://doi.org/10.1016/j.envhaz.2007.04.008.

within a model universal European 'Local Water Partnership' assigning functions to the Member States as well as to the organisation as a whole.

The study identifies cognitive gaps covering water conservation, and associated retention These are linked to:

– the implementation of a single, universal model for water management at local level across the European Union in support of the European Green Deal, cohesion policy, the common agricultural policy, regional policy and social policy.

– the implementation of a single European model for water conservation and retention based on local water conservation and retention strategies,

– creating the legal and financial conditions necessary for local water partnerships planning, developing and implementing local strategies,

– the protection of water resources in the form of legal and organisational solutions combining flood and drought prevention in a water management system integrating these two natural phenomena at local level,

– the allocation of financial resources (private, social, public national and community) to mitigate droughts and floods, prevent environmental pollution and large-scale fires, as part of a Europe-wide model for water retention and conservation.

The result of the analysis undertaken in this paper is to identify ways for the European Union to proceed in a post-pandemic environment. The approach is based on the concept of water conservation, and water retention, developed as part of the preliminary studies for the European Union's New Green Deal. This also included the projection of the organisation's development on the basis of a model of water conservation, water retention and, above all, climate-friendly agriculture. In addition, adapting and building the climate resilience of the organisation, its member states, capturing potential additional mitigation benefits, and sustainably boosting food security. Considerations include sustainable development, and social cohesion. The interpretation created in the study is intended to enable the realisation of climate smart agriculture. It does not resolve the dilemmas inherent in the nature of the Green New Deal, in particular limiting the impact of agriculture on environmental degradation, social inequality, labour exploitation, or the renewal of a post-pandemic capitalist economy. Nor was the question of what form of food consumption in Europe should ultimately be adopted resolved, and no reference was made to the associated environmental protection measures. The food consumption model – 'from the field' of the farmer 'to the table' of the consumer – was ignored in the discussion. References to the green closed-loop economy were not found in the work. Furthermore, there are marginal issues related to the creation of 'green' societies in European Union countries. The work undertook:

– verification of awareness of how to integrate water conservation into local water use planning using a universal methodology for building local water partnerships across the European community,

- to create an interpretation in which planning and management of water conservation at local level will be the responsibility of a partnership consisting of local government (depending on the administrative structure of each member state), water communities (depending on the legal and organisational form in each member state), farmer's/forest owners, social organisations, and entrepreneurs with organisational and financial support from a regional, state, and EU level,
- the creation of an interpretation of the methodology for developing local water conservation strategies using the LEADER method in their planning,
- to create an interpretation in which the coordination of basic preventive, protective, retention measures in the field of water management at local level (corresponding to the area of micro-catchments in the catchment area defined for the relevant water regions) to be integrated into the crisis management system of individual Member States and the EU as a whole,
- to point the way towards an adaptation of the water conservation planning system at local level with elements of coordination at regional and individual EU Member State level and across the whole organisation.

Empirical studies verifying the assumptions presented were not undertaken and carried out in this paper. Most importantly:

- testing the application of the adopted interpretation of building local water partnerships addressing climate risks (floods and droughts) on a European, national, regional, community scale,
- identifying and characterising the geological and hydrological characteristics of potential micro-basins and catchments of European rivers in terms of the applicability of methodologies for planning local water conservation strategies in individual countries, regions and communities,
- determining the level of investment required to implement local water conservation strategies in individual Member States as well as in the EU as a whole,
- reviewing the land-use plans of communities in the EU to ensure that they incorporate local water conservation strategies,
- comparative solutions with other countries and regions of the world,
- verifying the assumptions and application of the Local Water Partnership model in correlation with the crisis management systems of individual Member States,
- the formula for support from the European Union and its institutions to the proposed interpretation for local water partnerships,
- financial inputs and outputs in the projection of EU economic development policies and strategies (in the 5, 10, 15-, 20-, 25- and 50-year timeframes) for agriculture, forestry, fisheries, tourism and related economic sectors in the Member States,
- evaluating the multidirectional activities carried out to date by public and private entities, and state institutions in the area of water conservation effectiveness, which go beyond the scope outlined in the study.

The constraints indicated above indicate the direction of knowledge generation, accompanying solutions, and thus the expansion of the scope of interpretation presented in the study of the above-mentioned areas, and their offshoots.

The study assumes that national operational security systems of European Union Member States, tasked with water management, will take into consideration: 1) crisis management systems, 2) flood protection systems, 3) drought prevention systems, 4) water conservation systems. Additionally, complementary to 5) environmental protection systems, 6) fire protection systems. It was also assumed that local water conservation planning involving local water partnerships would be consistent with national crisis management systems, including flood protection and drought management. This will create the conditions for:

- water retention through the construction and upgrading of water storage reservoirs,
- the advancement of irrigation and drainage systems,
- construction and operation of flood protection structures,
- water flow control and retention,
- warning of hazardous phenomena in the atmosphere and hydrosphere,
- shaping the spatial development of river valleys and floodplains, building and maintaining dykes and other hydro technical installations,
- the exploitation of the Internet of Things in the activities listed above, providing additional opportunities to have environmental, economic, and security impacts.

The study assumes that the combination of the factors listed in the study into a synergistic whole will enable a comprehensive treatment of local critical infrastructure, including storage reservoirs and drainage and hydrological facilities at micro-catchment/catchment level. Moreover, their integration into local and regional development plans to:

- build a comprehensive water management system across all member states,
- acquire real-time information through the use of IT tools to support information management processes, such as those for water retention, flood protection, pollution detection, population alerting, and fire prevention,
- implementation of Economy 5.0 in terms of the Internet of Things (management, control, supervision, and administration of critical infrastructure elements).
- environmental protection through the development of water conservation involving a local water partnership within the micro-catchment / catchment area.

The current state of water conservation levels in the European Union is related to pressures on surface water bodies, hydro morphological pressures, diffuse sources of pollution particularly from agriculture and atmospheric deposition, and water abstraction. Coordinated grassroots water conservation efforts across the EU will ensure that a combination of local water conservation strategies developed and implemented by local water partnerships in conjunction with national emergency management systems will allow:

- harmonisation of legislation in the interconnected areas of human, environmental and economic security throughout the European Union,
- achieve synergies of normative, organisational, and operational relationships, and thus the overarching goal of reducing the risk of catastrophic droughts, floods, environmental contamination,
- coordination of the construction of critical infrastructure in this sphere for the protection of water resources and the prevention of the effects of natural disasters through the planned construction of retention and flood control measures.

The Local Water Partnership model shows ways to address the weaknesses and limitations of current EU activities identified. Aiming to achieve coherence in terms of organisation, function and competence within the security systems of individual Member States as well as the organisation as a whole.

The local water protection strategy is configured in three synergistically interlinked ways: flood protection, drought prevention, and pollution control. The aim of this monograph is to present, justify and develop the author's concept of a European / Community-wide universal model for water retention and conservation applicable across the European Union first defined in the work of Gromadzki Implementing the objectives of the European Green Deal, "Sustainable development and the European Green Deal vectors on the way to improving the scientist's workshop", Collective work edited by M. Staniszewski and H.A. Kretek, Silesian University of Technology Publishing House, Gliwice 2021., consistent with the model developed for Poland defined in the work Gryz, J., Gromadzki, S., Przeciwdziałanie suszy, Model of a European "Local Water Partnership". Implementing the objectives of the European Green Deal, "Sustainable development and the European Green Deal vectors on the way to improving the scientist's workshop", Collective work edited by M. Staniszewski and H.A. Kretek, Silesian University of Technology Publishing House, Gliwice 2021., consistent with the model developed for Poland defined in the study Gryz, J., Gromadzki, S., Przeciwdziałanie suszy Retencja wody w systemie zarządzania kryzysowego Polski, Wydawnictwo Naukowe PWN, 2021. Specific research objectives include:

1. Defining the significance of water conservation and retention, in the European Green Deal Strategy, including the identification of:
   - the importance of the interrelationship of water conservation and retention, in the context of dynamic climate change and the increasing threat of its effects in the form of drought, flooding, diminishing water resources, and water quality for agriculture and the economy of the European Union.
   - the weaknesses of the Union's strategy in terms of how it harnesses the strategic resource of water in its vision, objectives, priorities, actions and plans,
   - the current lack of a vision and proposal for a universal model of water conservation, and retention, for the entire Community,
   - how to plan and manage water conservation, retention, enhancing the competitiveness of the European Union,

     – the actual shape of the relationship between water conservation, and retention, in agricultural activities and environmental protection in the European Union.

2.   Defining European water resource management principles in line with the Water Framework Directive, based on a problem analysis of selected Member States including:
- proving the lack of a universal model for water resource management,
- the lack of a link between water retention and flood risks,
- the lack of a proposal for an effective, grassroots planning model for water resource protection, and retention, based on micro-catchments and the local community as the primary actors in the planning and management of water resource protection, and retention.

3.   The development and presentation of the concept of a universal European / Community model of a "Local Water Partnership", including:
- an analysis of what the EU has achieved so far in terms of grassroots development planning based on the LEADER programme and the so-called Local Action Groups (LAGs) operating in rural areas, and an indication of the possibility of using this model by extending the area of implementation, i.e., rural areas as defined in the Common Agricultural Policy, to micro / small catchment areas including urban areas,
- the development of a substantive and functional core body in the form of a "Local Water Partnership" planning, implementing and managing water protection, and retention, based on a local water protection and retention strategy,
- developing the basis of a methodology for establishing a "Local Water Partnership".

4.   The development and presentation of the concept of a universal European / Community model of a "Local Water Partnership", in terms of:
- an analysis of what the EU has achieved so far in terms of bottom-up development planning based on the LEADER programme and the so-called Local Development Strategies implemented in rural areas (indicating the possibility of using this model with an extension of the area of implementation to micro and small catchment areas including urban areas.
- development of the substantive and organisational principles for a "Local Strategy for Water Retention and Protection" as the basic planning document at local level, serving as a basis for the formulation of regional, national and Community strategies and plans,
- developing the basis of a methodology for developing a "Local Water Retention and Conservation Strategy".

The applicatory and implementation nature of the work will be achieved through:
- To bring the conceptual assumptions of the presented model to the scientific community and EU 'lawmakers' for debate.

- To attempt to integrate the insights presented in the work into discussions at the level of the European Parliament, the European Commission, the European Council and the governments of the Member States of the European Union.

The study was undertaken to answer the question of what water resource protection and retention model should be implemented in the European Union's policies in order to ensure the Union's effective ability to shape social cohesion, as well as economic, ecological (biodiversity conservation and restoration) and climate change security? The specific questions are:

- *How can the agricultural, industrial and environmental sectors contribute to making Europe more innovative?*
- *How to reinforce decarbonisation processes in the European Union using the agricultural, industrial and environmental sectors?*
- *How to shape the European Union's financial architecture recognising the needs for change in terms of climate protection, agriculture, sustainable development?*
- *In what direction to pursue work on technologies related to agriculture, sustainable social economy, and environmental protection that strengthen the competitiveness of these sectors?*
- *How to strengthen the competitiveness of the European Union in the areas of agriculture, industry and the environment?*
- *How to strengthen European Union social and economic recovery from the SARS COVID'19 pandemic leveraging on the agricultural, industrial and environmental sectors?*

The working proposition is that by implementing the proposed solutions contained in the model for the protection and retention of water resources, water retention will protect the environment, contribute to economic development by strengthening the agricultural sector, industry, innovation (economy 5.0) and also the security of the European Union. The supporting proposition is, that the existing organisational, functional, and normative solutions allow for a reconfiguration of the European Union's activities, policies and strategies, in order to meet the climate challenges of natural disasters, floods and droughts, within the existing crisis management formula of the organisation and that of the Member States.

The implementation of a universal model for water conservation, retention, across the European Union will provide additional opportunities for a sustainable CO_2 absorption policy. This model is designed to reinforce an EU-wide climate change adaptation strategy, fulfilling the spirit of the principles of subsidiarity and proportionality rights enshrined in the European Union's Charter of Fundamental Rights. It will also further contribute to enhancing experience and ensuring the systematic exchange of best practices on how to adapt to climate change. It will also enable organisational practices to be updated and aligned with the time horizon target of reducing net greenhouse gas emissions by at least 55% by 2030 with respect to 1990.

Foundations of future water resources management model

The basics of water management and safety standards related to it are defined by the regulations of national (internal) and international law. In the latter case, the leading role on the scale of the planet is played by the United Nations, and on the European continent by the European Union. The relationship of these two organizations defines the ways of thinking, political actions related to it, practices in various domains of water management, determining the nature of mutual influence between the water environment and the man who uses it.

Currently, it is imperative for the above-mentioned organizations to popularize access to water, to care for its resources and, where possible, to restore them and prevent degradation. These activities are comprehensive in nature, aimed at securing surface and subsurface waters used by humans. Actions aimed at preserving and renaturalizing the environment in order to increase biodiversity, which improves the quality of life of all creatures living on Earth, play an important role here.

The key to changing the existing formula of reality and creating a new, eco-empathetic formula with the participation of societies and states is the formula of the New Green Deal, which includes new solutions and policies that determine the nature of the use of water resources on Earth. Their importance is expressed in the concepts of green capitalism, which assumes the implementation of solutions that maintain a high quality of life for people on earth while actively protecting the natural environment. The context for the actions taken is the impact of man on the warming of the entire globe and the related changes in the natural environment.

1 Water security

The use of water resources has been associated with its protection, planned irrigation and safety for centuries. Today water security is defined by UN-Water as *"the capacity of a population to safeguard sustainable access to adequate quantities of acceptable quality water for sustaining livelihoods, human well-being, and socio-economic development, for ensuring protection against water-borne pollution and water-related disasters, and for preserving ecosystems in a climate of peace and political stability"*.[1] It embrace 10 components defining accesses and use of water in terms of drinking water, sanitation, good health, water quality and availability, water

---

1 Global water security 2023 assessment by administrator in UNU-INWEH reports, accessed 13.07.2023, https://inweh.unu.edu/global-water-security-2023-assessment/.

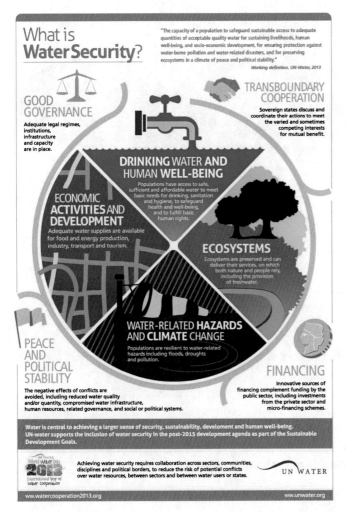

FIGURE 1.1
What is water security?
SOURCE: INFOGRAPHIC,
HTTPS://WWW.UNWA
TER.ORG/PUBLICA
TIONS/WHAT-WATER
-SECURITY-INFO
GRAPHIC, ACCESSED
14.07.2023

value, water governance, human safety, economic safety, water resource stability. These components are assessed and mapped at a national level using indicators with clear metrics and publicly available data. Where possible, single indicators are quantified using national SDG indicator data, freely available via online platforms maintained by UN SDG custodian data agencies. When this preliminary assessment was completed in early 2023, the most recent SDG indicator data available were for 2020, and unfortunately, over half of the water indicators had major data limitations that required the application of some sub-indicators and proxy values from open-source datasets.[2]

The UN approach introduces into practice four basic elements: 1. Good governance. 2. Transboundary cooperation. 3. Peace and political stability. 4. Financing (figure 1.1). All those elements bring different associations between states and

---

2 Ibid.

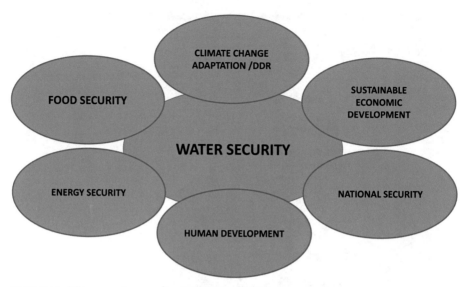

FIGURE 1.2 Water security according to the United Nations
SOURCE: OWN ELABORATION BASED ON UN APPROACH

international organizations. They combine water security with different forms of human activeness (figure 1.2).

Water security according to the UN is a multifaceted correlation between the impact of climate change on sustainable economic development, national security, human development, energy security and food security. From that perspective the UN and EU defines its approach towards water management as a tool of policy an international and national level of the states.

2   Water management: UN and EU approach

The phenomenon of water scarcity together with the increasing existential needs of the population, economic development and climate warming, progressing demographic growth together form the Gordian knot of humanity. These issues, individually and jointly, are the leading components of the policies pursued by the UN and its agencies, as well as by the European Union. At the same time, they are associated with the directions of change set by these organizations in terms of improving the existence of all humanity, as well as individual regions of our planet. The idea that accompanies the actions taken is related to changes in the production of food, its consumption, rational use of water resources and their restoration combined with the creation of conditions for the renewal of the natural environment. The correlation between man, the natural environment and its protection is obvious. Also obvious are remedial and anticipatory measures limiting the phenomenon of drought, excessive use of surface and subsurface waters, which should be introduced in the near future as part of Integrated Water Resources Management (IWRM).[3]

The basis for thinking about water resources by people on Earth is its use. Safe drinking-water, sanitation and hygiene are crucial to human health and well-being.[4] The UN trough 17 Sustainable Development Goals implements activities related to water – objective 6, Ensure availability and Sustainable management of water and sanitation for all.[5] The subject of the activities is the protection of water resources, its best possible quality and ways of using it by people.[6] It is directly related to the protection of the natural environment and man and is connected with green economy or green growth. The idea that inspires a variety of activities is inclusive green economy that improves human well-being, builds social equity and strive to reducing environmental risks and scarcities.[7] Green economy means the transition to more shared use resources, refers to a wide range of product-service, sharing and exchange practices that separate the ownership of the good from its use, while restoring social interaction and trust in the community of people with similar interests, "sharing" them.[8] This aspect can be regarded as leading in matters related to access to water, its retention, protection of the resource and its sharing.

---

3   Muhammad Mizanur Rahaman, Olli Varis, "Integrated water resources management: evolution, prospects and future challenges," Sustainability: science, practice and policy 1:1 (2005): 15–21.

4   Drinking unsafe water impairs health through illnesses such as diarrhoea, and untreated excreta contaminates groundwaters and surface waters used for drinking-water, irrigation, bathing and household purposes. Chemical contamination of water continues to pose a health burden, whether natural in origin such as arsenic and fluoride, or anthropogenic such as nitrate. Safe and sufficient WASH plays a key role in preventing numerous NTDs such as trachoma, soil-transmitted helminths and schistosomiasis. Diarrhoeal deaths as a result of inadequate WASH were reduced by half during the Millennium Development Goal (MDG) period (1990–2015), with the significant progress on water and sanitation provision playing a key role. Water, sanitation and hygiene (WASH), accessed 5.07.2023, https://www.who.int/health-topics/water-sanitation-and-hygiene-wash#tab=tab_1.

5   Ensure availability and sustainable management of water and sanitation for all, United Nations Department of Economic and Social Affairs Sustainable Development, accessed 5.07.2023, https://sdgs.un.org/goals/goal6.

6   Compendium of WHO and other UN guidance on health and environment, 2022 update. Geneva: World Health Organization; 2022 (WHO/HEP/ECH/EHD/22.01), accessed 6.07.2023, https://reliefweb.int/report/world/compendium-who-and-other-un-guidance-health-and-environment?gad_source=1&gclid=CjwKCAiA6byqBhAWEiwAnGCA4J_QCBlpYZOUcokwWJAmHKuvFHQXogXMEFgyAOozJ1MihJZTy2QONhoCP_YQAvD_BwE.

7   About green economy, UN environment programme, accessed 6.07.2023, https://www.unep.org/explore-topics/green-economy/about-green-economy.

8   Uncovering pathways towards an inclusive green economy a summary for leaders, United Nations Environment Programme, 2015, accessed 6.07.2023, https://wedocs.unep.org/bitstream/handle/20.500.11822/9838/-_Uncovering_Pathways_towards_an_Inclusive_Green_Economy_a_Summary_for_Leaders-2015IGE_NARRATIVE_SUMMARY_Web.pdf.pdf?sequence=3&amp%3BisAllowed=y%2C%20Portuguese%7C%7Chttps%3A//wedocs.unep.org/bitstream/handl.

Water scarcity and droughts on the European continent are becoming more frequent and widespread. There are currently two phenomena. The first is over-exploitation of available water resources, which exacerbates the effects of droughts. Second, the deterioration of water resources in Europe, as well as seasonal and long-term shortages. In 2019, 38% of the population of the European Union was affected by water restrictions, 29% of the territory of the Union was directly affected by drought.[9] The costs of this phenomenon amounted to over EUR 12 billion per year. Conservative, lower bound estimates show that exposing today's EU economy to global warming of 3 °C above pre-industrial levels. This would result in an annual loss of at least EUR 170 billion (1.36% of EU GD). Slow onset sea level rise is also an increasing worry for coastal areas, which produce ~40% of the EU GDP and are home to ~40% of its population.[10] In connection with this state of affairs, the European Union has adopted guidelines to mitigate and potentially reverse the effects of the described negative phenomena. These are guidance on water prices, water allocation, drought prevention and response, alternative water supplies, high-quality information and technology solutions to combat water scarcity and droughts.[11] Policies and actions of the EU are set up to prevent and to mitigate water scarcity and drought situations trough a water-efficient and water-saving economy.

Regulations are the basis for the European Union's activities in the field of water supply to agriculture. The first concerns the reuse of water. First of all, from properly treated wastewater, for example from municipal wastewater treatment plants, reducing its wastage by households, industry, implementing appropriate water acquisition systems. In this respect, pro-health policies related to food hygiene, application of standards, preventive measures (in transport, production, consumption of goods) are correlated. The result is to be the reuse of water for irrigation in agriculture as part of a circular economy.[12] These activities are correlated with other activities related to the use of water by humans, primarily related to risk assessment and risk management for catchment areas for water intake points intended for human consumption, which will be carried out for the first time by July 12, 2027. In addition, the assessment and risk management in the water supply system,

9   Water scarcity and droughts. Preventing and mitigating water scarcity and droughts in the EU, accessed 6.07.2023, https://environment.ec.europa.eu/topics/water/water-scarcity-and-droughts_en.

10   Communication from the Commission to the European Parliament, the Council, the European Economic and Social Committee and the Committee of the Regions empty forging a climate-resilient Europe – the new EU strategy on adaptation to climate change, European Commission, COM(2021) 82 final, Brussels, 24.2.2021.

11   Rozwiązanie problemu susz i niedoboru wody w UE, accessed 5.07.2023 https://eur-lex.europa.eu/PL/legal-content/summary/addressing-water-scarcity-and-droughts-in-the-eu.html#.

12   Regulation (EU) 2020/741 of the European Parliament and of the Council of 25 May 2020 on minimum requirements for water reuse, Official journal of the European Union, L 177/32, 5.6.2020.

which is to take place for the first time by 12 January 2029.[13] Considering the above, it becomes obvious that there is a need to correlate knowledge about water quality, quantity and access in micro-catchments and catchments, both for agriculture, the economy and the population due to the sharing of this resource. The need to protect it is articulated in the currently processed directives on wastewater and pollution, e.g. by industrial farms.[14]

The European Green Deal adopted by the European Commission and endorsed at EU level in December 2019 creates opportunity to accelerate water policy management in Europe. Work Programme 2022–2024 on Common Implementation Strategy EU WATER LAW integrates both water quality and water quantity into all relevant EU policies. The main objective of this 2022–2024 CIS Work program is to accompany, support and enable a much enhanced implementation of water legislation. It embraces enhanced cross-sectoral approach of the European Green Deal, the challenges of the water management. In particular, the *Farm to Fork Strategy* which contribute to reduce pressures coming from certain polluting farming practices. Then after the *Biodiversity Strategy for 2030* that emphasis on nature protection and restoration, i.e., the aquatic and marine environment and has been correlated with the *new Circular Economy Action Plan*. This plan promotes water reuse and efficiency, including in industry and refers to water reuse and circularity in urban wastewater treatment. Together with an Integrated Nutrient Management Action Plan, the *Chemicals Strategy for Sustainability*, the *Renovation Wave, the Pharmaceutical Strategy, Sustainable and Smart Mobility Strategy, the formula of action aims at the new Climate Adaptation Strategy which* steer the EU to a climate resilient society by 2050. Mentioned policies, strategies and out coming actions focuses on water quantity and finally on *Zero Pollution Action Plan for water, air and soil* to eliminate pollution as a pressure on our water bodies and lays out a program of actions to reduce pollution pressure.[15]

The actions in water protection according to the implementation of the Water Framework Directive entering into a crucial period. In 2027, it will no longer be possible to invoke the most applied exemptions for water bodies not in good status. Holistic approach entailed by the European Green Deal and related initiatives, with

---

13   Directive (EU) 2020/2184 of the European Parliament and of the Council of 16 December 2020 on the quality of water intended for human consumption, Official journal of the European Union, 23.12.2020.

14   Proposal for a Directive of the European Parliament and of the Council concerning urban wastewater treatment (recast), European Commission, COM(2022) 541 final, Brussels, 26.10.2022. Proposal for a Directive of the European Parliament and of the Council amending Directive 2010/75/EU of the European Parliament and of the Council of 24 November 2010 on industrial emissions (integrated pollution prevention and control) and Council Directive 1999/31/EC of 26 April 1999 on the landfill of waste, European Commission, COM(2022) 156 final/3 Strasbourg, 5.4.2022.

15   CIRCABC, CIS work programme 2022–2024.docx, accessed 6.07.2023, https://circabc.europa .eu/ui/group/9ab5926d-bed4-4322-9aa7-9964bbe8312d/library/561e8b77-e75d-42d6-86a9-164055 47735f/details.

a view to achieve, in the most effective manner, are focused on investments, valuation, incentives and pricing, implementation and compliance, integration with other policy areas (agriculture, energy, transport etc.), chemical pollution (including substances of emerging concern), administrative streamlining, monitoring and digitalization. New challenges in water management helps revamping efforts on water quantity management and water efficiency, with measures aiming at improving the planning coordination across sectors, promoting effective nature-based solutions, reducing climate-related risk and ensuring the availability and sustainability of fresh water. The European Commission focus on securing greater public awareness, in a manner conducive to actions at both EU and national/local level incentivizing water efficiency in buildings, from rainwater harvesting to enhanced flood protection via nature based solutions.[16]

Due to a changing climate, many regions of Europe are already facing more frequent, severe and prolonged droughts, which are already having cascading effects. As a result, they lower the water level in rivers and groundwater, inhibit the growth of trees and crops, increase pest attacks, and contribute to fires. In Europe, most of the losses caused by drought (around EUR 9 billion per year) are related to agriculture, the energy sector and public water supply. The extreme droughts in Western and Central Europe in 2018, 2019 and 2020 caused significant damage, which was mitigated by short-term emergency measures under the Union Civil Protection Mechanism. Currently, EU countries are implementing integrated river basin management through the Water Framework Directive. Some have adopted drought management plans for vulnerable river basins, but almost all of them may be vulnerable. Therefore, organizational and technical adaptation solutions are needed in agriculture, including sustainable water use, soil and vegetation management, implementation of drought-tolerant crops, vertical farming, strategic planning of land use and reclamation of damaged areas. In the energy and transport sectors, this includes e.g., prepare for disruptions on individual waterways with freight, hydropower and cooling for power plants. In addition, restrictions on drinking water supplies and the necessary promotion of water saving in residential buildings, additional water supply and storage infrastructure.[17]

An important role in above respect plays information's gathered trough the European Drought Observatory which is part of the Copernicus Emergency Management Service. The European Commission also launched the European Drought Observatory for Resilience and Adaptation Project (EDORA). The aim is to enhance drought resilience, adaptation and cooperation throughout the EU

---

16   Ibid.

17   Communication from the Commission to the European Parliament, the Council, the European Economic and Social Committee and the Committee of the Regions empty forging a climate-resilient Europe – the new EU strategy on adaptation to climate change, European Commission, COM(2021) 82 final, Brussels, 24.2.2021.

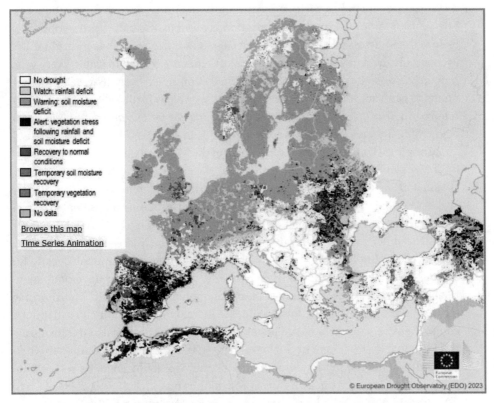

MAP 1.1   Situation of combined drought indicator in Europe – 2nd ten-day period of June 2023
SOURCE: EUROPEAN DROUGHT OBSERVATORY, ACCESSED 6.07.2023,
HTTPS://EDO.JRC.EC.EUROPA.EU/EDOV2/PHP/INDEX.PHP?ID=1000

states who are gaining the financial support from European Commission for different projects.

What is more in context of drought, extraordinary situation may also cause usage of the EU Civil Protection Mechanism established in 2001 by the European Commission.[18]

3   Legal formula of water resources management within UN

The United Nations' key activities for the protection of water resources, its retention and use are included in the resolutions. Those normative acts are supplemented in

---

18   In 2022, the Mechanism was activated 106 times to respond to war in Ukraine; wildfires in Europe; COVID-19 in Europe and worldwide, i.e. floods in Pakistan. EU Civil Protection Mechanism, accessed 3.07.2023, https://civil-protection-humanitarian-aid.ec.europa.eu/what/civil-protection/eu-civil-protection-mechanism_en.

the organization's policies with guidelines for directional actions in specific areas. The solutions adopted by the organization, as well as the policy, are periodically reviewed and assessed by the Secretary-General's office as well as individual UN agencies involved in various areas of activities that are related to water. Every year, they are assessed during a conference that comprehensively assesses progress, points to gaps, summarizes the actions taken, as well as further directions of work referred to *Transforming our world: the 2030 Agenda for Sustainable Development* adopted by General Assembly of United Nations in year 2015.[19]

During 2023 United Nations Conference on the Midterm Comprehensive Review of the Implementation of the Objectives of the International Decade for Action, "Water for Sustainable Development", 2018–2028, on March 22–24, 2023 in New York.[20] Within Interactive dialogue 5: Water Action Decade achievements in implementation were considered Agenda 21 Resolution adopted by the General Assembly on 21 December 2020 United Nations Conference on the Midterm Comprehensive Review of the Implementation of the Objectives of the International Decade for Action, "Water for Sustainable Development", 2018–2028. This document mentions, among other, resolution 70/1 of 25 September 2015, entitled "Transforming our world: the 2030 Agenda for Sustainable Development", in which it adopted a comprehensive, far-reaching and people-centered set of universal and transformative Sustainable Development Goals and targets. What is more, committed to working tirelessly for the full implementation of the Agenda by 2030, its recognition that eradicating poverty in all its forms and dimensions, including extreme poverty, is the greatest global challenge and an indispensable requirement for sustainable development.[21] According to this outline the sustainable development goals and targets related to water resources and sanitation, including those contained in the 2030 Agenda for Sustainable Development, and determined to achieve the goal of ensuring the availability and sustainable management of water and sanitation for all and other related goals and targets.[22] It was then assumed that water is critical for sustainable development and the eradication of poverty and hunger, that water,

---

19   Seventy-fifth session Agenda item (a) Sustainable development: towards the achievement of sustainable development: implementation of the 2030 Agenda for Sustainable Development, including through sustainable consumption and production, building on Agenda 21 Resolution adopted by the General Assembly on 21 December 2020 [on the report of the Second Committee (A/75/457/Add.1, para. 14)] 75/212. United Nations Conference on the Midterm Comprehensive Review of the Implementation of the Objectives of the International Decade for Action, "Water for Sustainable Development", 2018–2028.

20   2023 United Nations conference on the midterm comprehensive review of the implementation of the objectives of the international decade for action, "Water for sustainable development", 2018–2028, United Nations, A/CONF.240/2023/8, 31 January 2023.

21   Resolution adopted by the General Assembly on 25 September 2015 transforming our world: the 2030 Agenda for Sustainable Development, General Assembly, United Nations, A/RES/70/1, 21 October 2015.

22   The United Nations world water development report 2019: leaving no one behind, United Nations Educational, Scientific and Cultural Organization, Paris (2019): 1–34.

ecosystems, energy, food security and nutrition are linked and that water is indispensable for health, well-being and human development, including the empowerment of women, and a vital element of achieving the Sustainable Development Goals and other relevant goals in the social, environmental and economic fields.[23]

During 2023 United Nations Conference on the Midterm Comprehensive Review of the Implementation of the Objectives of the International Decade for Action, "Water for Sustainable Development", 2018–2028 was considered the actions were formulated in the UN Secretary-General's Plan: Water Action Decade 2018–2028. They implement the strategy of social, ecological and economic development of the planet in the area of access to water. They focused on:

1. Advance sustainable development. Today, unsustainable development pathways and governance failures negatively affect the quality and availability of water resources, compromising their capacity to generate social and economic benefits and increasing disaster risk. Driven by growing demands from manufacturing, thermal electricity generation and domestic use, global water demand is projected to increase by 55% by 2050. The Decade aims to bring a greater focus to sustainable development and a risk-informed integrated management of water resources for the achievement of social, economic and environmental objectives.

2. Energize implementation of existing programs and projects. A concerted effort water-related challenges of UN Member States, the UN system, civil society and the private sector to improve access to safe water and sanitation, reduce pressure on water resources and ecosystems, manage water-related disaster risks, adapt to climate change, reduce water pollution and increase reuse in constantly needed. The actual Decade provides a unique opportunity to energize on-going activities of all relevant actors through improved cooperation, partnership and capacity development.

3. Mobilize action to achieve the 2030 Agenda, a roadmap to a better, more sustainable world with water. Guaranteeing sustainable water management is a vital element of achieving the Sustainable Development Goals and other relevant goals in the social, environmental and economic fields. One of the goals, no. 6, addresses the issues relating to drinking water, sanitation and hygiene, the quality and sustainability of water resources worldwide interlinkages to all other 16 Goals, of which a majority are mutually reinforcing.[24]

---

23 In earlier documents indicated the outcome of the document of the high-level panel on water, entitled "Making every drop count: an agenda for water action", the *Sustainable development goal 6 synthesis report on water and sanitation*, the outcomes and Ministerial Declaration of the eighth World Water Forum, held in Brasilia from 18 to 23 March 2018, the outcomes of the United Nations special thematic sessions on water and disasters, the outcomes of the Budapest Water Summit in 2019 and the Sustainable Development Goal 6 Global Acceleration Framework.

24 The facilitate action, the objectives of the Decade are pursued through four work streams: a) Facilitating access to knowledge and the exchange of good practices; b) Improving knowledge generation and dissemination; c) Pursuing advocacy, networking and promoting partnerships;

In this context the 2022 Secretary-General's Report on the midterm review of the Decade highlighted accomplishments and pinpointed some key actions, events, and lessons learned to make a success and help achieve SDG 6.[25]

During the UN 2023 Water Conference the voluntary commitments of Member States and relevant stakeholders provided an avenue through Water Action Agenda and explored how the outcomes of the previous events and initiatives at the midway point can accelerate the progress. Achievements are not satisfying. Unfortunately, the world has not come closer to achieving SDG 6 – water and sanitation for all and related goals by 2030. The hydrological cycle that connects and supports terrestrial, freshwater and marine ecosystems does not provide the assumed positive cultural, environmental, economic and political values. The COVID-19 pandemic has highlighted the hitherto unresolved links between water, its management and the three pillars of sustainable development. In addition, the need to build resilience, especially among the planet's most vulnerable communities. Water is a key determinant of achieving the internationally agreed goals in 2030 Agenda for Sustainable Development, the 2015 Paris Agreement and the Sendai Framework for Disaster Risk Reduction 2015–2030.[26] The UN Secretary-General's Plan on the Water Action Decade 2018–2028 recognizes water as being at the heart of these recent agreements.

---

d) Strengthening communication for implementation of the water-related SDGs. United Nations Secretary-General's Plan: Water Action Decade 2018–2028, accessed 7.04.2023, https://sdgs.un.org/sites/default/files/2021-05/UN-SG-Action-Plan_Water-Action-Decade-web_0.pdf.

25   The SDG 6, Ensure availability and sustainable management of water and sanitation for all, includes sub goals. 6.1. By 2030 indicates achieving universal and equitable access to safe and affordable drinking water for all. 6.2. By 2030, achieving access to adequate and equitable sanitation and hygiene for all and end open defecation, paying special attention to the needs of women and girls and those in vulnerable situations 6.3. By 2030, improving water quality by reducing pollution, eliminating dumping and minimizing release of hazardous chemicals and materials, halving the proportion of untreated wastewater and substantially increasing recycling and safe reuse globally. 6.4. By 2030, substantially increasing water-use efficiency across all sectors and ensure sustainable withdrawals and supply of freshwater to address water scarcity and substantially reduce the number of people suffering from water scarcity. 6.5. By 2030, implementing integrated water resources management at all levels, including through transboundary cooperation as appropriate. 6.6 By 2020, protecting and restoring water-related ecosystems, including mountains, forests, wetlands, rivers, aquifers and lakes. 6.a By 2030, expanding international cooperation and capacity-building support to developing countries in water- and sanitation-related activities and programmes, including water harvesting, desalination, water efficiency, wastewater treatment, recycling and reuse technologies. 6.b Supporting and strengthen the participation of local communities in improving water and sanitation management. Resolution adopted by the General Assembly on 25 September 2015 Transforming our world: the 2030 Agenda for Sustainable Development, General Assembly, United Nations, A/RES/70/1, 21 October 2015: 18–19.

26   Sendai framework for disaster risk reduction 2015–2030, (UNISDR: Switzerland 2015), 12–37.

The recently adopted 2022 Kunming-Montreal Global Biodiversity Framework replacing the Aichi Targets also recognizes the role of water. The main concerns focus on increasing climate extremes and variability, coupled with unsustainable growth and consumption. All together is leading to more severe and frequent water-related disasters and risks, worsening environmental degradation including pollution, increasing water temperatures and ecosystem loss, and profoundly affecting economies, societies and the environment. This, in turn, undermines the natural capacity of ecosystems to combat both the causes and effects of climate change. Projections indicate that the increase in global warming will increase the risk to ecosystems and people. In the last decade, nine out of ten disasters were caused by water-related natural hazards. Dependence on water makes food security, human health, urban and rural settlements, energy production, industrial development, economic development, ecosystems more and more dependent on water, and water on the effects of climate change. Water resources and hydrological processes are subject to climate change processes.[27]

The United Nations World Water Development Report 2020, Water and Climate Change, states that climate change, unsustainable human activities and environmental mismanagement affect the availability, quality and quantity of water, impeding the realization of the human right to water and sanitation, a clean and healthy environment and other related human rights. In 2018, 2.3 billion people (almost 30% of the global population) lived in countries under water stress and 3.6 billion people faced inadequate access to water at least one month per year. Building resilience refers to building the capacity of a system, community or society at risk to resist, absorb, accommodate, adapt. In addition, transforming and restoring the effects of the hazard in a timely and effective manner, including by preserving and restoring its essential structures and functions through risk management. Events such as *SARS COVID-19*, devastating floods, show that many countries lack the necessary preparation, coping capacity and management systems to deal with the systemic nature of the risk. The result is the likelihood of natural disaster hazards affecting large areas and population centers. This can be related to the cascading effects of natural disasters. Water-related risks are growing at an unprecedented rate, as are the frequency, intensity and costs of natural disasters. This causes significant losses and damage to people, nature, economic assets and infrastructure. The death toll from water-related disasters has more than doubled in the last 10 years. Nearly 95% of losses and damage to infrastructure reported between 2010 and 2019 were caused by water disasters. During this period, around 1.4 billion people were affected by droughts and 1.6 billion by floods between 2000 and 2019. As climate impacts do not recognize borders, and 60% of global freshwater supplies, are

---

27   Kunming-Montreal Global Biodiversity Framework, conference of the parties to the convention on biological diversity, CBD/COP/DEC/15/4, 19 December 2022, 4–15.

found in transboundary basins shared by 153 countries, this adds an international dimension to climate change adaptation and disaster risk reduction. Greenhouse gas (GHG) emissions also come from water-based processes. Wastewater treatment plants and sludge disposal methods tend to generate methane, which is a very potent greenhouse gas. Sustainable water management could help avoid and reduce carbon, methane and nitrous oxide emissions from water and wastewater management as well as poorly managed or drained freshwater systems such as peatlands.[28]

The Intergovernmental Panel on Climate Change (IPCC) in their 2022 report projects an increase of water-related hazards and threats to water availability and quality with and exacerbated by increased global warming. This affects agricultural and energy sectors, ecosystem integrity as well as river basins dependent on snowmelt, glaciers, groundwater availability and surface water storage. The IPCC projects that an increase of global warming to 2 or 3°C can cause direct flood damages that are 1.4 to 3.9 times higher than the 1.5°C global warming scenario without adaptation. Moreover, even though most documented climate change adaptation measures respond to water-related risks and impacts, their effectiveness is hampered by increased global warming.[29] Global temperatures continue to rise unabated and cause climate extremes and related disasters including water-related ones. In that respect countries have inadequate risk knowledge as a key bottleneck to strengthening early warning systems. The risk data ecosystem needs to be strengthened through better risk analyses and tracking of losses and damages to be able to manage water-related disasters. Measuring, gathering knowledge, risk-informing decisions, and tracking progress in resilience building, are critical to sustainable development.[30] UN 2023 Water Conference stressed that data sharing, both within and among countries, needs to be promoted. Monitoring of implementation and efforts to build resilience should be enhanced through reporting mechanisms of SDGs, Sendai Framework and related initiatives. Political processes should be informed by science and evidence.[31] There is a strong need to adequately address

---

28   UNESCO, UN-Water, 2020: United Nations world water development report 2020: water and climate change, (Paris: UNESCO), 10–181.

29   Hans-Otto Pörtner, Debra C. Roberts, Melinda Tignor, Elvira Poloczanska, Katja Mintenbeck, Anders Alegría, Marlies Craig, Stefanie Langsdorf, Sina Löschke, Vincent Möller, Andrew Okem, Bardhyl Rama (eds.), IPCC, 2022: Climate change 2022: impacts, adaptation and vulnerability. contribution of working Group II to the sixth assessment report of the intergovernmental panel on climate change, (Cambridge University Press, Cambridge, UK and New York, NY, USA), 3056, doi:10.1017/9781009325844.

30   WMO provisional state of the global climate 2022, World Meteorological Organization, accessed 12.04.2023, https://library.wmo.int/doc_num.php?explnum_id=11359.

31   Data and information on groundwater are particularly lacking as it is difficult to generate, but groundwater is a vitally important source, providing around half of the world's total drinking water and support ecosystems. 66 Investments in data on groundwater dynamic, volume and identification of recharge areas at national level are critical. 2023 United Nations Conference

climate, resilience and environmental challenges related to water protection and use. The complex and interdependent challenges of climate change, disaster risk reduction and environmental degradation and their impacts on water demands cross-cutting theme across all SDG 6 accelerators. Institutional and human capacity development that is inclusive, enables innovation, including the use of artificial intelligence, virtual reality and digital learning, and new forms of collaboration have to be exanimated.[32]

An important element of whole water-based puzzle refers to green jobs and a skilled workforce of water professionals. UN agencies strongly encourages citizens and open science to address climate and resilience challenges, mainly to improve water management. Nature-based Solutions range from native vegetation rather than concrete to control soil erosion and reduce water runoff. They can also be an effective way to address some of our key societal challenges. What is more, enhance biodiversity while also maintaining and creating employment and improving labor productivity. Nature-based Solutions create needs for innovation. Mainly through Investments in solutions and technologies that can help to better manage water resources and facilitate both adaptation to and mitigation against climate change. Only integrated solutions that tackle climate change adaptation and mitigation and help overcome systemic risks, while providing environmental, social and economic benefits enabling conditions as well as stimulate innovation supported by enabling policies and regulations. This includes knowledge-sharing and uptake by making the business case for local water management solutions between countries. The communities affected by climate change need to feel ownership over their own water management such that they are able to design workable and sustainable solutions that incorporate their experiences and knowledge in building resilience. They can help elevate the role of water resources in climate mitigation and for disaster risk reduction as well as promote innovative and alternative solutions, such as nature – based or hybrid solutions and circular economy.[33]

4   Legal formula of water resources management within EU

The normative basis for the management of water resources and its protection in the European Union are created by Articles 191–192 of the *Treaty on the Functioning of the European Union* referring to the environment. The Article 191 imposes:

---

on the Midterm Comprehensive Review of the Implementation of the Objectives of the International Decade for Action …

32   Early warnings for all the UN global early warning initiative for the implementation of climate adaptation, executive action plan 2023–2027, World Meteorological Organization, accessed 12.04.2023, https://www.preventionweb.net/media/84612/download.

33   2023 United Nations conference on the midterm comprehensive review of the implementation of the objectives of the international decade for action …

1. Union policy on the environment shall contribute to pursuit of the following objectives:
   - preserving, protecting and improving the quality of the environment,
   - protecting human health,
   - prudent and rational utilization of natural resources,
   - promoting measures at international level to deal with regional or world-wide environmental problems, and in particular combating climate change.
2. Union policy on the environment shall aim at a high level of protection taking into account the diversity of situations in the various regions of the Union. It shall be based on the precautionary principle and on the principles that preventive action should be taken, that environmental damage should as a priority be rectified at source and that the polluter should pay. In this context, harmonization measures answering environmental protection requirements shall include, where appropriate, a safeguard clause allowing Member States to take provisional measures, for non-economic environmental reasons, subject to a procedure of inspection by the Union.
3. In preparing its policy on the environment, the Union shall take account of:
   - available scientific and technical data,
   - environmental conditions in the various regions of the Union,
   - the potential benefits and costs of action or lack of action,
   - the economic and social development of the Union as a whole and the balanced development of its regions.
4. Within their respective spheres of competence, the Union and the Member States shall cooperate with third countries and with the competent international organizations. The arrangements for Union cooperation may be the subject of agreements between the Union and the third parties concerned.

In above respect the Article 192 includes:

1. The European Parliament and the Council, acting in accordance with the ordinary legislative procedure and after consulting the Economic and Social Committee and the Committee of the Regions, shall decide what action is to be taken by the Union in order to achieve the objectives referred to in Article 191.
2. By way of derogation from the decision-making procedure provided for in paragraph 1 and without prejudice to Article 114, the Council acting unanimously in accordance with a special legislative procedure and after consulting the European Parliament, the Economic and Social Committee and the Committee of the Regions, shall adopt:
   (a) provisions primarily of a fiscal nature.
   (b) measures affecting:
      - town and country planning,
      - quantitative management of water resources or affecting, directly or indirectly, the availability of those resources,
      - land use, with the exception of waste management.

(c)   measures significantly affecting a Member State's choice between different energy sources and the general structure of its energy supply.
The Council, acting unanimously on a proposal from the Commission
and after consulting the European Parliament, the Economic and Social
Committee and the Committee of the Regions, may make the ordinary
legislative procedure applicable to the matters referred to in the first
subparagraph.

3.   General action programs setting out priority objectives to be attained shall be
adopted by the European Parliament and the Council, acting in accordance
with the ordinary legislative procedure and after consulting the Economic
and Social Committee and the Committee of the Regions.[34]

On the basis of the treaty the directives are adopted jointly by the European
Parliament and the Council of the European Union. They are addressed to all
Member States after publication in the Official Journal of the European Union.
They enter into force on the date indicated by them or on the twentieth day from
the announcement. The method of achieving the objectives set in them depends on
individual Member States, their internal legal acts. Directives are the basis for the
implementation of the organization's policy. The European Commission is responsible for monitoring the implementation of the directives and their supervision
related to their application. Changes in directives over time are conditioned by the
strategies and detailed policies of the European Commission, the executive body of
the European Union.

The basis for the action of the European Union for the retention and protection
of water resources is Directive 2000/60/EC of the European Parliament and of the
Council of 23 October 2000 establishing a framework for Community action in the
field of water policy (named shortly The Water Framework Directive). It establishes
a framework for the protection of inland surface waters, transitional waters, coastal
waters and groundwater that:

–   prevent further deterioration, protect and enhance aquatic ecosystems and, with
regard to their water needs, terrestrial ecosystems and wetlands directly dependent on aquatic ecosystems.
–   promote sustainable water use based on long-term protection of available water
resources.
–   seek increased protection and improvement of the aquatic environment, including through specific measures to progressively reduce discharges, emissions and
losses of priority substances and to cease or phase out discharges, emissions and
losses of priority hazardous substances.

---

34   Consolidated version of the treaty on the functioning of the european union, part three:
union policies and internal actions – title XX: environment, Official journal 115, 09/05/2008,
0132–0133.

- ensure a gradual reduction of groundwater pollution and prevent its further pollution,
- contribute to reducing the effects of floods and droughts, and thus contribute to ensuring an adequate supply of good quality surface and groundwater, which is essential for sustainable and equitable water use.[35]

In order to achieve this action in the field of water policy Annex VII River Basin Management Plans refers to achieve the water protection objectives set out in the policy documents by adopting, EU-wide method for the European Green Deal.[36] The annex, in formula of report, analysed Danube basin throughout the whole Romania, impact of the Action Program for the protection of waters against pollution caused by nitrates from agricultural sources and of the Code of good agricultural practice. This was an attempt to recognize Pan-European model of collecting data and delivering actions.

The Directive 2000/60/EC of the European Parliament and of the Council of 23 October 2000 establishing a framework for Community action in the field of water policy is supplemented by more detailed legal acts issued in different periods of time. These are:

- Council Directive of 12 June 1986 on the protection of the environment, and in particular of the soil, when sewage sludge is used in agriculture.
- Council Directive 91/271/EEC of 21 May 1991 concerning urban waste-water treatment and Commission Directive 98/15/EC of 27 February 1998 amending Council Directive 91/271/EEC with respect to certain requirements established in Annex I thereof.
- Council Directive 91/676/EEC of 12 December 1991 concerning the protection of waters against pollution caused by nitrates from agricultural sources.
- Council Directive 98/83/EC of 3 November 1998 on the quality of water intended for human consumption.

---

35   The aim is to take actions of states that will contribute to a significant reduction of groundwater pollution, protection of territorial and sea waters, protection and prevention of marine environment pollution by cessation or gradual elimination of discharges, emissions of hazardous substances. The ultimate goal to be achieved in the marine environment is to achieve concentrations close to background values for naturally occurring substances and close to zero for man-made synthetic substances. Directive 2000/60/EC of the European Parliament and of the Council of 23 October 2000 establishing a framework for Community action in the field of water policy, Official journal of the European Communities 22.12.2000.

36   This annex report is one product of the Study on European Union (EU) integrated policy assessment for the freshwater and marine environment, on the economic benefits of EU water policy and on the costs of its non-implementation" (BLUE2) commissioned by the European Commission (EC) (Luxembourg: Publications Office of the European Union, 2019), 7–26. Annex VII. Application of the bottom-up multicriteria methodology in eight European River Basin Districts – The Arges-Vedea RBD. Deliverable to Task A3 of the BLUE 2 project "Study on EU integrated policy assessment for the freshwater and marine environment, on the economic benefits of EU water policy and on the costs of its nonimplementation". Report to DG ENV.

– Directive 2006/7/EC of the European Parliament and of the Council of 15 February 2006 concerning the management of bathing water quality and repealing Directive 76/160/EEC.
– Directive 2006/118/EC of the European Parliament and of the Council of 12 December 2006 on the protection of groundwater against pollution and deterioration.
– Directive 2007/60/EC of the European Parliament and of the Council of 23 October 2007 on the assessment and management of flood risks.
– Directive 2008/56/EC of the European Parliament and of the Council of 17 June 2008 establishing a framework for community action in the field of marine environmental policy (Marine Strategy Framework Directive).
– Directive 2008/1/EC of the European Parliament and of the Council of 15 January 2008 concerning integrated pollution prevention and control.
– Directive 2008/105/EC of the European Parliament and of the Council of 16 December 2008 on environmental quality standards in the field of water policy, amending and subsequently repealing Council Directives 82/176/EEC, 83/513/EEC, 84/156/EEC, 84/491/EEC, 86/280/EEC and amending Directive 2000/60/EC of the European Parliament and of the Council.
– Directive (EU) 2020/2184 of the European Parliament and of the Council of 16 December 2020 on the quality of water intended for human consumption.

All together, they establish formula of actions undertaken by European Commission and EU Member States both on land and the sea.

The European Union, based on the above-mentioned directives, conducts a water policy expressed in EU 2030 Biodiversity Strategy. If focus on efforts to restore freshwater ecosystems and the natural functions of rivers. In this respect the proposed universal model of water resources management within water security partly fulfils the strategy's agenda. The target of restoring rivers to a free-flowing state is designed to support and find synergies between efforts to achieve the Water Framework Directive objectives and the EU Birds and habitats Directives, with the overarching aim of boosting the restoration of freshwater ecosystems.[37] One of the leading aims is preventing and mitigating water scarcity and droughts in the EU to ensure access to good quality water in sufficient quantity for all Europeans, economic sectors and the environment. What is more, to ensure the good status of all water bodies across Europe.[38] It has been pointed out by Union's ability to respond to the increasing pressures on water resources could be improved by wider reuse of treated waste water, limiting extraction from surface water bodies and groundwater

---

37 European Commission, Directorate-General for the Environment, *2030 biodiversity strategy: removing barriers to restoring rivers*, Publications office of the European Union, 2022, accessed 18.04.2023, https://data.europa.eu/doi/10.2779/858614.

38 Water scarcity and droughts, accessed 18.04.2023, https://environment.ec.europa.eu/topics/water/water-scarcity-and-droughts_en.

bodies, reducing the impact of discharge of treated waste water into water bodies, and promoting water savings through multiple uses for urban waste water, while ensuring a high level of environmental protection.[39] In that respect there is an urgent need to create an instrument to regulate standards at Union level for water reuse, in order to remove the obstacles to a widespread use of such an alternative water supply option. This can help to reduce water scarcity and lessen the vulnerability of supply systems.[40]

Specific role in water protection within EU plays integrated water management set in 2008. The formula of "Water information notes" gives an introduction and overview of key aspects of the implementation of the Water Framework Directive.[41] It embraces twelve actions described in Water Notes. Water Note 1 focuses on Joining Forces for Europe's Shared Waters, coordination in international river basin districts that cover the territory of more than one Member State and for coordination of work in these districts.[42] Water Note 2 introduce cleaning up Europe's Waters. It is identifying and assessing surface water bodies at risk. The Water Framework Directive improving the status of many water bodies across the EU and it establishes principles for water management, i.e. including public participation in planning and economic approaches, including the recovery of the cost of water services.[43] Water Note 3 Groundwater at Risk, managing the water under us focuses on ensuring a stable quantity of clean groundwater body which cannot be exceed to the rate at which freshwater replenishes it. This has implications for coastal areas, where seawater can seep into freshwater aquifers and spoil precious freshwater resources.[44] Water Note 4 Reservoirs, Canals and Ports: Managing artificial and heavily modified water bodies reflects treatment of physically altered rivers and other waters for navigation, flood control, barge canals and hydroelectric reservoirs and other

---

39   Regulation (EU) 2020/741 of the European Parliament and of the Council of 25 May 2020 on minimum requirements for water reuse.

40   Communication from the Commission to the European Parliament, the Council, the European Economic and Social Committee and the Committee of the Regions a blueprint to safeguard europe's water resources, Brussels, 14.11.2012 COM(2012) 673 final.

41   Water notes – about integrated water management, EU water legislation and the Water Framework Directive, accessed 18.04.2023, https://ec.europa.eu/environment/water/participation /notes_en.htm.

42   Water note 1 joining forces for europe's shared waters: Coordination in international river basin districts, European Commission (DG Environment), March 2008, accessed 18.04.2023, https://ec.europa.eu/environment/water/participation/pdf/waternotes/water_note1_joining _forces.pdf.

43   Water note 2 cleaning up Europe's waters: identifying and assessing surface water bodies at risk, European Commission (DG Environment) March 2008, accessed 18.04.2023, https://ec .europa.eu/environment/water/participation/pdf/waternotes/water_note2_cleaning_up.pdf.

44   Water note 3 groundwater at risk. managing the water under us, european commission (DG Environment) March 2008, accessed 19.04.2023, https://ec.europa.eu/environment/water/par ticipation/pdf/waternotes/water_note3_groundwateratrisk.pdf.

purpose created in the places where no water bodies previously existed.[45] Water Note 5 Economics in Water Policy: The value of Europe's waters rises up environmental costs including damage to ecosystems such as pollution that harms fish and wildlife in rivers. Extracting water for human causes repercussions such as reducing water levels in rivers and lakes, and this may harm ecosystems. Preferred approach includes i.e. estimates of costs and benefits of existing water supply systems and the investment costs of new water supply or wastewater treatment systems as well as indirect benefits such as an increase in jobs if cleaner coastal waters lead to higher tourism levels should also be considered.[46] Water Note 6 Monitoring programs: taking the pulse of Europe's waters embraces long-term surveillance, operational monitoring, investigative monitoring which tracks the effectiveness of investments and other measures taken to improve the status of water bodies.[47] Water Note 7 Intercalibration: A common scale for Europe's waters sets the basic requirements for measuring the health of surface water ecosystems. It identifies four common "quality elements" to be used in determining ecological status: phytoplankton; other aquatic flora; benthic (bottom dwelling) invertebrate fauna and fish fauna.[48] Water Note 8 Pollution: Reducing dangerous chemicals in Europe's waters focusing on emissions, discharges and losses, combining of control measures, establishing limits, known as Environmental Quality Standards (EQS), polluting of different substances (all together 41 of them).[49] Water Note 9 on the Implementation of the Water Framework Directive Integrating water policy linking all EU water legislation within a single framework based on Water Framework Directive.[50] Water Note 10 on the Implementation of the Water Framework Directive Climate change: Addressing floods, droughts and changing aquatic ecosystems reflects the impact of climate change for the northern and southern part of the continent. It points out the role of

---

45   Water note 4 reservoirs, canals and ports: managing artificial and heavily modified water bodies, European Commission (DG Environment) March 2008, accessed 19.04.2023, https://ec.europa.eu/environment/water/participation/pdf/waternotes/water_note4_reservoirs.pdf.

46   Water note 5 economics in water policy: the value of Europe's waters, Commission (DG Environment) March 2008, accessed 19.04.2023, https://ec.europa.eu/environment/water/participation/pdf/waternotes/water_note5_economics.pdf.

47   Water note 6 monitoring programmes: taking the pulse of Europe's waters, Commission (DG Environment) March 2008, accessed 19.04.2023, https://ec.europa.eu/environment/water/participation/pdf/waternotes/water_note6_monitoring_programmes.pdf.

48   Water note 7 intercalibration: a common scale for Europe's waters, Commission (DG Environment) March 2008, accessed 19.04.2023, https://ec.europa.eu/environment/water/participation/pdf/waternotes/water_note7_intercalibration.pdf.

49   Water note 8 pollution: reducing dangerous chemicals in Europe's waters, Commission (DG Environment) December 2008, accessed 19.04.2023, https://ec.europa.eu/environment/water/participation/pdf/waternotes/water_note8_chemical_pollution.pdf.

50   Water note 9 water notes on the implementation of the Water Framework Directive integrating water policy: linking all EU water legislation within a single framework, Commission (DG Environment), December 2008, accessed 20.04.2023, https://ec.europa.eu/environment/water/participation/pdf/waternotes/water_note9_other_water_legislation.pdf.

planning ahead in each river basin, which Member States can address through their river basin management plans (RBMPs).[51] Water Note 11 on the Implementation of the Water Framework Directive From the rivers to the sea: Linking with the new Marine Strategy Framework Directive.[52] An ecosystem approach has been adopted ensuring good environmental status and involving protecting of marine ecosystems, their measurement.[53] Water Note 12 A Common Task: Public Participation in River Basin Management Planning includes calls for the public to be informed and involved in the preparation of river basin management plans, which identify measures to improve water quality.[54]

All, mentioned above, Water Notes was playing crucial role in forging formula of water protection within EU serving as a basis for future activeness. At the same time, between 2012 and 2020 as an initiative within the EU 2020 Innovation Union the European Innovation Partnership on Water (EIP Water) was established. The aim of this initiative was to:

- speed up innovations that contribute to solving societal challenges,
- enhance Europe's competitiveness and contribute to job creation and economic growth,
- help to pool expertise and resources by bringing together public and private actors at EU,
- develop innovative solutions to address major European and global water challenges,
- support the creation of market opportunities for innovations, both inside and outside of Europe.[55]

In this regard, the EU has set its objectives. The Strategic Implementation Plan was treated as a milestone for defining EU priorities across a broad group of stakeholders

51 Water note 10 water notes on the implementation of the Water Framework Directive climate change: addressing floods, droughts and changing aquatic ecosystems, Commission (DG Environment) December 2008, https://ec.europa.eu/environment/water/participation /pdf/waternotes/water_note10_climate_change_floods_droughts.pdf, accessed 20.04.2023.

52 Water note 11 water notes on the implementation of the Water Framework Directive from the rivers to the sea: linking with the new Marine Strategy Framework Directive, Commission (DG Environment) December 2008, https://ec.europa.eu/environment/water/participation /pdf/waternotes/water_note11_marine_strategy.pdf, accessed 20.04.2023.

53 Commission Decision (EU) 2017/848 of 17 May 2017 laying down criteria and methodological standards on good environmental status of marine waters and specifications and standardised methods for monitoring and assessment, and repealing Decision 2010/477/EU. Report from the Commission to the European Parliament and the Council on the implementation of the Marine Strategy Framework Directive (Directive 2008/56/EC), Brussels, 25.6.2020.

54 Water note 12 a common task: public participation in river basin management planning, Commission (DG Environment) December 2008, accessed 20.04.2023, https://ec.europa .eu/environment/water/participation/pdf/waternotes/water_note12_public_participation _plans.pdf.

55 EIP Water website, accessed 20.04.2023, https://ec.europa.eu/environment/water/innovation partnership/index_en.htm.

included in the Steering Group.[56] Especially in bringing together actors from across the water sector, Increasing the visibility of the water sector amongst EU policy-makers, facilitating networking amongst water sector stakeholders. In consequence 29 voluntary, multi-stakeholder Action Groups were established, e.g., developing an anaerobic membrane reactor for water treatment, improving the water management implementation capacities of cities and regions, building solutions for coastal aquifer monitoring and management, rolling-out a benchmarking process for the entire water cycle, consolidating the sector of nature-based technologies, promoting multi-stakeholder river contracts. Finally, the main achievements of the EIP Water made ground for EU Biodiversity Strategy for 2030, the "Farm to fork strategy", the Circular Economy Action Plan and the Zero pollution action plan. Those achievements were aligned with the objectives of water legislation, as well as with the European Green Deal presented in 2019.

An immanent element of the whole water protection construction is permanent access to the data trough Water Information System for Europe (WISE). It creates an information gateway to water issues in two areas. First one focus on freshwater information system for Europe.[57] Within three domains the web side brings information about:

- Country profiles on urban wastewater treatment. This section presents section presents key data related to the implementation of the Urban Wastewater Treatment Directive (UWWTD) in Europe. Country profiles are available for each EU-27 Member State, as well as for Norway and Iceland. The European Union profile presents similar information, aggregated to EU-27 level.[58]
- WISE freshwater resource catalogue which shows all kind of information related to water treatment in human action profile.[59]

---

56   The EIP Water conferences brought together the various actors in Brussels (2013), Barcelona (2014), Leeuwarden (2016), Porto (2017) and Zaragoza (2019) to showcase technologies and innovation, visualize the progress of Action Groups, explore market opportunities, discuss the barriers which hamper innovation uptake in practice, and present policy approaches to improve and strengthen innovation in the EU water sector. Collaborative processes for change and innovation within EIP Water embraced public and private sector, non-governmental organizations and the general public. At that time 29 voluntary, multi-stakeholder Action Groups were established developing an anaerobic membrane reactor for water treatment, improving the water management implementation capacities of cities and regions, building solutions for coastal aquifer monitoring and management, rolling-out a benchmarking process for the entire water cycle, consolidating the sector of nature-based technologies, promoting multi-stakeholder river contracts. Ibid.

57   Source. https://water.europa.eu/freshwater, accessed 24.04.2023.

58   Country profiles on urban waste water treatment, accessed 24.04.2023, https://water.europa.eu /freshwater/countries/uwwt.

59   WISE freshwater resource catalogue, accessed 24.04.2023, https://water.europa.eu/freshwater /data-maps-and-tools/metadata.

- Data, maps and facts collected for purpose of "reporting obligations" that means collection and reporting data as a basis for the assessment of the European environmental status and of the policies' effectiveness.[60]

Second element included in WISE is marine system information for Europe focusing on:

- MSFD Reporting Data Explorer, the tool dedicated to Marine experts. It provides information on the main Central Data Repository of the Eionet website, where the Member States upload files from 2012 up to date with the information requested on each of the articles.[61]
- Visualization tools on Good Environmental Status (GES) assessments reported by the EU Member States to the European Commission of the current environmental status of their marine waters. Results on the Good Environmental Status assessments under Article 8 are presented in a dynamic format.[62]
- Dashboards on marine features under other policies which present the ecological and chemical status assessments under the Water Framework Directive, the conservation status of marine species and habitats under the Habitats Directive and the coverage of Marine Protected Areas in European waters.[63]
- Country Profile Factsheets which include information on the marine environment of EU Member States, including the state of features reported under different Directives.[64]

All together gatherer information of water use within European Union and plays supplementary role in creation ambitions, directions and finally political goals, strategies and directives, legal acts of EU Institutions and its Member States.

5   The legal outline for the European New Green Deal

A start point for a European Union New Green Deal was published in March 2009 United Nations Environment Programme (UNEP) which issued a major policy brief, Global Green New Deal. It included: climate action, environmental rights

---

60   Data, maps and tools, accessed 24.04.2023, https://water.europa.eu/freshwater/data-maps-and
      -tools.

61   MSFD reporting data explorer, accessed 24.04.2023, https://water.europa.eu/marine/data
      -maps-and-tools/msfd-reporting-information-products/msfd-reporting-data-explorer.

62   GES assessments dashboards, accessed 24.04.2023, https://water.europa.eu/marine/data-maps
      -and-tools/msfd-reporting-information-products/ges-assessment-dashboards.

63   Dashboards on marine features under other policies, accessed 24.04.2023, https://water.europa
      .eu/marine/data-maps-and-tools/map-viewers-visualization-tools/dashboards-on-marine
      -features-under-other-policies.

64   Country profile factsheets, accessed 24.04.2023, https://water.europa.eu/marine/countries-and
      -regional-seas/country-profiles.

and governance, environment under review.[65] The idea was to coordinate various national economic stimulus plans in a manner closely coupled enough to become new worldwide formula of policy. First of all, in the way of consuming goods, the formulas for conducting a green policy in this area. The Green Deal has found its justification as a form of individual and social action in stopping global warming, the accompanying climate and environmental changes, retraining the workforce and counteracting unemployment, combating poverty, improving the existing infrastructure, reinvesting capital and developing technology.[66] The Green Deal, as a projection of social and political expectations gaining strength, also pointed to benefits, financial and health, individual and collective, for societies.[67] The promoted approach redefined the environment in terms of costs, benefits, assets, liabilities, gains, losses and values.[68] In addition, during this period of time, the implementation of appropriate solutions was indicated that will enable the implementation of a viable eco-policy in the current environmental crisis.[69]

The vision of the Green New Deal, promoted since the beginning of the 21st century, highlights the need for a holistic and cross-sectoral approach, whereby all relevant national policies contribute to achieving the overarching climate goal of reducing pollution and its removal by the natural environment while sustaining consumption.[70] The formula of the changes includes initiatives in a number of closely related areas, such as climate, environment, energy, transport, industry, agriculture and sustainable finance.[71] Currently, the European Green Deal is a package of policy initiatives aimed at putting the EU on the path of ecological transformation and ultimately achieving climate neutrality by 2050. I am going to transform the union into an entity in which a fair, prosperous society with a modern, competitive economy develops. The adaptation of policies, strategies, directives, and standards of the European Union to the concept of implementing the New Green Deal has been taking place since the 2000s of the 21st century. At the same time,

---

65   Global Green New Deal Policy Brief, United Nations Environment Programme, March 2009, 1–6, 8–9.

66   John Robert McNeill, *Something new under the sun: an environmental history of the twentieth century world.* (New York: Norton, 2000), 3–21, 118–150.

67   David Bach, Hilary Rosner, *Go green, live rich: 50 simple ways to save the earth (and get rich trying)*, (New York: Broadway Books, 2008), 2–7.

68   Ernst U. von Weizacker, Amory B. Lovins, L Hunter Lovins, *Factor four: doubling wealth – halving resource use: a report to the Club of Rome*, (New York: Kogan Page 2001), 271–277.

69   Timothy W. Luke, "A green new deal: why green, how new, and what is the deal?", Critical policy studies, 3:1 (2009): 14–28, DOI: 10.1080/19460170903158065.

70   Timothy W. Luke, "Situating knowledges, spatializing communities, sizing contradictions: the politics of globality, locality and green statism," in *Environmental governance: power and knowledge in a local–global world*, ed. Gabriela Kütting, Ronnie Lipschitz (New York: Routledge, 2009), 13–37.

71   David Powell, Lukasz Krebel, Frank van Lerven, Five ways to fund a Green New Deal. We can afford it. We can't afford not to, New economics foundation, 28 November 2019, accessed 30.04.2023, https://neweconomics.org/2019/11/five-ways-to-fund-a-green-new-deal.

it was a response to the political challenges associated with many contemporary environmental crises.

The foundations of the concept of the European Union's Green Deal should be sought in 2010 in a document adopted by the European Commission. It was EUROPE 2020 a Strategy for smart, Sustainable and inclusive growth, which indicated three priorities: smart development – development of an economy based on knowledge and innovation; diversified development – supporting an efficient economy that uses natural resources, is more environmentally friendly and more competitive; inclusive growth – supporting a high-employment economy that ensures social and territorial cohesion.[72] The adopted document was a kind of supplement to the aforementioned strategy taken on 13 February 2012, by the European Commission, an outline for Innovating for Sustainable Growth: A Bioeconomy for Europe. This document embedded proposes for a comprehensive approach to ecological, environmental, energy, food supply and natural resource challenges of modern Europe and the world. It spells out what the Bioeconomy is and a roadmap to lay the foundations for a more innovative, resource-efficient, competitive society where food security is compatible with the sustainable use of renewable resources for industrial purposes, while protecting the environment.[73]

On 20 June 2019 the European Council adopted conclusions on the next institutional cycle. They included climate change, disinformation and hybrid threats, external relations, the European Semester and enlargement. The European Council notice the importance of the United Nations Secretary General's Climate Action Summit for stepping up global climate action so as to achieve the objective of the Paris Agreement (to limit the temperature increase to 1.5 °C above pre-industrial levels).[74] On 11 December 2019, the European Commission published its communication a European Green Deal Striving to be the first climate-neutral continent implementing transformation of the EU into a modern, resource-efficient and competitive economy, ensuring: no net emissions of greenhouse gases by 2050, economic growth decoupled from resource use, no person and no place left behind (till 2023 more than 40 actions where adopted in accordance with strategy).[75] On 12 December 2019 EU leaders endorsed the objective of making the EU climate-neutral by 2050, in accordance with the Paris Agreement. They underlined

---

72   Communication from the Commission EUROPE 2020 a strategy for smart, sustainable and inclusive growth, Brussels, 3.3.2010 COM (2010) 2020, 3–32.

73   Innovating for sustainable growth a bioeconomy for Europe, European Commission, Directorate-General for Research and Innovation 2012, 8–48.

74   European Council meeting (20 June 2019) – conclusions, General Secretariat of the Council, EUCO 9/19.

75   European Green Deal striving to be the first climate-neutral continent, European Commission, https://commission.europa.eu/strategy-and-policy/priorities-2019-2024/european-green -deal_en, accessed 30.04.2023.

that the transition to climate neutrality will bring significant opportunities for economic growth, markets, jobs and technological development.[76]

An important moment in the context of the New Green Deal was the adoption on March 5, 2020, by the Council the EU's submission to the United Nations Framework Convention on Climate Change (UNFCCC) on the long-term low greenhouse gas emission development strategy of the EU and its member states. In this statement the EU and its member states committed to the Paris Agreement and its long-term goals, i.e., ambitious social and economic transformation, aiming to inspire global climate action moving towards climate neutrality. Another crucial document was adopted on 20 May 2020, EU Biodiversity Strategy for 2030. It included comprehensive, ambitious and long-term plan to protect nature and reverse the degradation of ecosystems with aims to build societies' resilience to future threats such as the impacts of climate change, forest fires, food insecurity, disease outbreaks including by protecting wildlife and fighting illegal wildlife trade.[77]

The enforcing role for European Green Deal have been undertaken by EU a Farm to Fork Strategy for a fair, healthy and environmentally – friendly food system. The document focuses on priorities related to the food chain and sets out how to make Europe the first climate-neutral continent by 2050. It maps a new, sustainable and inclusive growth based on robust and resilient food system. Farm to Fork Strategy creates a new comprehensive approach to how Europeans value food sustainability at the same time improving lifestyles, health, and the environment. In details the document creates directions of how to reduce dependency on pesticides and anti-microbials, reduce excess fertilization, increase organic farming, improve animal welfare, and reverse biodiversity loss.[78] Today, the 'farm to fork' strategy is a roadmap to build a sustainable European Union food system, in line with the aims of the European Green Deal.[79]

An important role in the spectrum of actions undertaken by EU was adoption of the document on 11 March 2020 A new Circular Economy Action Plan for a cleaner and more competitive Europe. It creates one of the main building blocks of the European Green Deal, Europe's new agenda for sustainable growth and jobs to achieve the EU's 2050 climate neutrality target and to halt biodiversity loss. The Circular Economy Action Plan imposes initiatives along the entire life cycle of

---

76   European Council meeting (12 December 2019) – conclusions, General Secretariat of the Council, EUCO 29/19.

77   Biodiversity strategy for 2030, European Commission Environment, accessed 30.04.2023, https://environment.ec.europa.eu/strategy/biodiversity-strategy-2030_en.

78   Communication from the Commission to the European Parliament, the Council, the European Economic and Social Committee and the Committee of the Regions a farm to fork strategy for a fair, healthy and environmentally-friendly food system, COM/2020/381 final.

79   At a glance taking the EU's 'farm to fork' strategy forward, European Parliament, https://www .europarl.europa.eu/RegData/etudes/ATAG/2021/690622/EPRS_ATA(2021)690622_EN.pdf, accessed 30.04.2023.

products with economic actors, consumers, citizens and civil society organizations. Foremost, it targets how products are designed, promotes circular economy processes, encourages sustainable consumption, and aims to ensure that waste is prevented, and the resources used are kept in the EU economy for as long as possible.[80]

Finally, the European Union New Green Deal focuses on Pathway to a Healthy Planet for All EU Action Plan: Towards Zero Pollution for Air, Water and Soil. The fight against pollution has been claimed as a fight for fairness and equality due to the fact it threatens biodiversity and significantly contributes to the on-going mass extinction of species. The EU zero pollution ambition is a cross-cutting objective contributing to the UN 2030 Agenda for Sustainable Development. It is also complementing the 2050 climate-neutrality goal in synergy with the clean and circular economy and restored biodiversity goals being part and parcel of many European Green Deal and other initiatives.[81]

6 · Conclusions

The formula of the Green Deal, proposed in 2008 by the United Nations and adopted in 2019 by the European Union for implementation, indicates the key determinants constituting the choices of humanity in the 21st century. Firstly, relating to water retention in the natural environment, changing the formula of the economy of the planet, a single farmer and agriculture as a sector affecting the condition of the natural environment. Secondly, the successive implementation of solutions that would concern the activity of states and international organizations would create conditions for zero-emission production and consumption of agricultural and other goods. Thirdly, they would focus on creating a system of water protection, surface and subsurface, in rural and urbanized areas. The indicated group of determinants is related to multi-directional activities undertaken by the UN and the EU, which focus on stopping negative climate change, making countries and societies more resilient, guided by the principles of responsibility and interdependence, the imperative of not excluding individuals, social groups, countries or regions of our planets. The synergy of the activities of the UN and the EU is obvious, the long-term results of these activities are not obvious, which depends on the will and policy of the states gathered in both organizations.

---

80   Communication from the Commission to the European Parliament, the Council, the European Economic and Social Committee and the Committee of the Regions a new circular economy action plan for a cleaner and more competitive Europe, Brussels, 11.3.2020 COM(2020) 98 final.

81   Communication from the Commission to the European Parliament, the Council, the European Economic and Social Committee and the Committee of the regions Pathway to a healthy planet for all EU action pPlan: 'Towards zero pollution for air, water and soil', Brussels, 12.5.2021 COM(2021) 400 final.

The formula for the development of the modern world requires the adaptation of existing forms of human activity to the needs of protecting the natural environment and the planet. This is dictated by the imperative action of all humanity in the form of preserving the natural environment, its biodiversity, in this and future generations. This creates demands for the more specific policies of European Union. One of them has been articulated by Water Europe which is promoter of water-related innovation and RTD in Europe and advocates it. This action group has an idea to set out a blueprint for a society in which the true value of water is recognized and realized. What is more all available water sources are managed in such a way that water scarcity and pollution of water are avoided, water and resource loops are largely closed to foster a circular economy and optimal resource efficiency. At the same time the water system is resilient against the impact of climate change events.[82] The aspect of damages, costs and benefits related to adaptation to climate change creates the basis for the development and implementation of a new universal model of protection of water resources and its retention within the proposed local water partnership. In the case of the European Union, it is about supporting the adaptation of Member States in key vulnerable sectors of their economies. In addition, adapting the common agricultural policy to climate threats while ensuring a more resilient social infrastructure.

The relationship of economic benefits and losses, human activity in the processes of production and consumption of goods, a specific lifestyle, has a constant impact on nature and water management. It is usually negative and expressed by increasing the amount of its consumption, as well as the devastation of this natural resource in the form of discarded waste: plastics, as well as pollutants such as heavy metals and many other substances harmful to animals, people, and the entire natural environment. The need to reduce environmental pressure by man and his actions is obvious. Therefore, the need to reduce human pressure on the natural environment, retention and protection of water resources becomes a necessity. In the activities carried out by the UN and the EU, there are constantly growing needs in the form of the assessment of anthropogenic pollution, assessed by the size of the load flowing into the water reservoir, water quality and stability of water resources, maintaining or restoring the good condition of the ecosystem, and reacting in the event of various types of negative phenomena. The need to refine this comprehensive approach is evident. The question that remains open at the moment is how to approach this, taking into account the different climatic conditions on our planet.

---

82   WE is a membership-based multi-stakeholder organisation representing over 200 members from academia, industry, technology providers, water users, water service providers, civil society, and public authorities. Zero pollution action plan: a first step towards a blue deal for a water-smart society, water Europe technology & innovation, https://watereurope.eu/wp -content/uploads/Position-paper-Zero-Pollution-Strategy-FINAL-1.pdf.

When considering the issue of the possibility of actions taken by the UN and the EU in the formula of the Green Deal, it should be stated that adequate normative foundations were developed in the first two decades of the 21st century. The scope of policies of organizations and states based on them remains adequate to the identified issues of climate change, their implications for the natural environment, societies and their economy. It is still an open question how to effectively conduct activities that change the negative trends of human impact on nature that emerged in the past centuries. The projection of protection of water resources, its retention appropriate to given climatic and environmental conditions, is becoming the central issue of the 21st century humanity. Nowadays, aquatic, marine and terrestrial ecosystems are an important component of the environment. Land ecosystems perform water retention functions in the form of swamps, lakes, water reservoirs, rivers, canals, fishponds and ditches, which unfortunately undergo degradation processes as a result of not always rational human activity.

Local water partnership as a new universal model of water resources management within water security is one of the possible, untested solutions. In principle, it creates an opportunity to improve the water management system by creating its integrated form, based on two complementary pillars: preventing floods and droughts by increasing the level of security in the area of water management, food management, biodiversity protection, preventing ecological threats and the involvement of farmers, entrepreneurs, local communities, social organizations, ecologists, public institutions into joint activities. In addition, the use of innovative technical potential, existing and dedicated to specific conditions for the protection of water resources, retention and solutions.

CHAPTER 2

Determinants of water retention in Poland

Conservation of water resources in the European Union is part of the global climate puzzle. Its puzzle depends both on what is done 'here and now' and the impact it will have on the decades and centuries to come. Considering the projection of planet-wide changes, the 'Polish climate puzzle' may seem insignificant, but it represents a picture of the whole. Furthermore, they condition the existence and development of society between the Bug and Nysa Łużycka rivers, the Baltic Sea and the Carpathian and Sudeten mountains. Their impact is not only significant in the Central European Plain, but also affects other surrounding areas.

The projection of weather change will lead to an increasing adoption of technology and the use of the energy and infrastructure necessary for this. In this respect, water retention, and drought prevention in the management system of the state and the European Union, creates an opportunity of developing superior instruments in each area of human activity, starting with security. With their identical ecological, economic and security goals in mind, together they form a framework for synergistic action.

At present, water retention and drought prevention in Poland are linked to making it possible to live in areas that are subject to steppification (in the future even desertification) or that in any other way require the use of appropriate measures to improve climate conditions, the natural environment for plants, animals and people.[1] Moreover, the adaptation and use of technology for the 5.0 economy, and to strengthen national security. However, this requires a re-evaluation of the way we approach climate security issues in Poland and more broadly the European continent.

1   Poland's natural environment: a problematic situation

Today's efforts to reduce the greenhouse effect, assuming that the reduction of greenhouse gas emissions to "zero" would be successful, would be at least two decades before any result is achieved, considering that this would happen in 2017. This would not take place until after 2035. The warming of the oceans, the reduction of the ice surface around the Arctic, the retreat of glaciers, and the rise in the waters

---

1   Paul C.D. Milly, K.A. Dunne, *Projected percentage changes in runoff, 21st century. An ensemble of 12 climate models participating in the 3rd phase of the coupled model intercomparison project*, Geophysical fluid dynamics laboratory, Princeton University, accessed 10.08.2020, https://www.gfdl.noaa.gov/wp-content/uploads/pix/user_images/pcm/runoff_change_animation.gif.

of the seas and oceans would continue as at present.[2] In this respect, therefore, we are referring to ideas and not to current realities. This means that the human environment will continue to change. In the future, agriculture will be dependent on efficient water supply systems. Open questions remain as to whether this is achieved in a way that is environmentally predatory (water abstraction resulting in a lowering of the water table and subsurface water levels, as is the case in many regions of the world) or sustainable, with an emphasis on restoring the environment with the help of water. Projecting its development over time will depend on the answers that are given in Poland. Their result can contribute to positive changes not only along the Vistula, but also in other regions of the European continent and even the planet. This is about the efforts that Poland and the European Union could make if the proposed transformations related to water retention, and drought prevention, are successful here.

The broadening of the spatial, climatic and ecological scope of national security means that Poland is increasingly looking for a framework to integrate its own environmental security into the broader structures of climate governance. The warming of the Earth's climate is not a uniform process. However, it is progressing most rapidly in the northern hemisphere. Changes in this region of the planet are caused by the melting of Arctic glaciers. The consequence is a shifting of the zone suitable for cultivation as well as for habitation into increasingly northern areas of the planet. This situation favours the position of Poland, as well as the countries of the European Union, with regard to the development of agriculture in the 21st century. It also potentially contributes to the value of this sector of the economy.[3] However, it will depend on the measures taken, right here and now, whether Poland, as well as the European Union, will take advantage of this opportunity, or whether the state, with the tacit consent of the European Union in which there is no systemic approach to solving the problems of water retention and drought prevention, will allow the devastation of the natural environment as a result of agricultural activity without taking into account the outlined climatic trend. Paradoxically, water retention and drought prevention in the crisis management system, whether of Poland or the European Union, create an opportunity to meet – at least in part – the climate challenges of the XXI century. This is because it favours the sustainable development of agriculture by addressing the key resource that determines its health and development prospects – water.

The United Nations Intergovernmental Panel on Climate Change issued a report in October 2018 on the effects of a global warming of 1.5 °C. This is in comparison to its pre-industrial levels, as well as the reductions in global greenhouse gas emissions undertaken. Currently, man-made global warming has now reached 1 °C

---

2 *Strategic foresight analysis 2017 report,* Allied commander transformation NATO, 67.
3 Peter Ferguson, "The green economy agenda: business as usual or transformational discourse?," Environmental politics 24 (2015): 17–37, https://doi.org/10.1080/09644016.2014.919748.

above pre-industrial levels. It is increasing at an average rate of around 0.2 °C per decade during the XXI century. Without appropriate international climate action, the average global temperature increase could reach 2 °C shortly after 2060 and continue to rise thereafter.[4] As the European Commission's Clean Planet for All Communication indicates, this is expected to have serious consequences for the productivity of the European economy, infrastructure, and food production capacity. Also, for public health and biodiversity, political stability.

In recent years, large areas of southern and western Europe have suffered from the effects of severe droughts. In the case of Central and Eastern Europe, areas of it have been especially hard hit by floods in recent years. In 2017, weather-related disasters caused economic losses of EUR 283 billion, which were experienced by 5% of Europe's population. By 2100, projections predict, around two-thirds of the continent's population could be affected. The annual damage caused by river flooding in Europe could rise from the current EUR 5 billion to EUR 112 billion.[5] The consequences could be catastrophic, similarly to what happened in 1997 in southern and western Poland, in the Czech Republic, eastern Germany, north-western Slovakia, and eastern Austria.[6] Here, too, it is appropriate to make use of water retention, which, by being stored in a number of micro reservoirs, can in principle lead to a targeted improvement in the water balance on a local and nationwide scale. Costs, therefore, can be significantly reduced for reconstruction and restoration of the natural environment.

With reference to Poland, the provisions of the European Commission's Communication A Clean Planet for All … then two elements under consideration are to be expected, i.e., a reduction in precipitation and its abrupt, sudden form. Thus, the state is exposed to two distinct atmospheric phenomena, which it can and should effectively counter. Once again, it should be emphasised that the Clean Planet for All … document identifies eight scenarios for changing the modifying the identified trend. **None of them address water retention, or drought mitigation as part of crisis management, or security management at the local level or macro level – the European Union Member State or the organisation as a whole.** They all focus on reducing greenhouse gas emissions either directly or indirectly.

---

4   Communication from the Commission "A clean planet for all", The European Commission strategic long-term vision for a prosperous, modern, competitive and climate-neutral economy, European Commission, Brussels, 28.11.2018, COM(2018) 773 final.

5   Ibid.

6   In Poland, "*The following rivers flooded out: Nysa Łużycka, Nysa Kłodzka, Oder, Widawa and the upper Vistula. The most intensive rain fell from 3 to 10 July. The disaster affected 25 provinces at the time. The water flooded nearly 700,000 hectares and 1,362 villages, and also some large cities. Nearly 200 000 people were evacuated. (…) All the losses were estimated by the government of Włodzimierz Cimoszewicz at 9 billion zlotys, but it later transpired that they amounted to 12 billion. The floods left 7,000 people homeless. The floodwater destroyed or damaged 680,000 homes, 4,000 bridges, 613 kilometres of dykes and 500,000 hectares of farmland.*" Polskie Radio PL, Portal Polskiego Radia, accessed 11.08.2020, audio, https://www.polskieradio.pl/39/248/Artykul/633313,Powodz-tysiaclecia.

It should therefore be noted that the European Union does not have specific ideas for combining flood prevention with drought prevention. Far too many European Commission initiatives result in fragmented policies in this area.[7] This is primarily due to the defined profile of climate investment in the natural environment by Member States, i.e., the elimination of CO_2 from the atmosphere through its absorption, the purification of water, and the protection of biodiversity in the natural environment.

The above discussion defines a new potential direction for the European Union and, at the same time, a paradigm for thinking about organisational security. Considering the ambitious goals of the organisation, such as the "exercise leadership in global climate action", the solutions identified in the study can serve as a reference point for the organisation's policies and strategies.[8] All the more so as the EU has specialised instruments at its disposal, such as the European Fund for Strategic Investment, cohesion policy funds and the European Agricultural Fund for Rural Development (EAFRD), which can easily be adapted to the needs of water resource conservation throughout the EU, albeit with different effects in individual regions. European Investment in Rural Areas, with the European Social Fund and the Reconstruction Fund under the Recovery Plan for Europe (after the Coronavirus pandemic), which could easily be adapted (additionally prioritised) to address water retention needs across the community albeit with varying effects in different regions.

With reference to Poland, in the Programme Assumptions for Counteracting Water Deficit for 2021–2027 with a perspective to 2030 adopted by the government, a direction complementary to the described approach to crisis management

---

7 Joanne Linnerooth-Bayer, Anna Dubel, Jan Sendzimir, Stefan Hochrainer-Stigler, *Challenges for mainstreaming climate change into EU flood and drought policy: water retention measures in the Warta River Basin, Poland*, Regional environmental change 15, no. 6 (2014): 1017–1023, DOI:10.1007/s10113-014-0643-7.

8 Current information on the European Union's initiatives in the field of comprehensive climate protection is as follows:
1.   The financial sector in the support of the climate.
2.   The European external investment plan – opportunities for Africa and the EU neighbourhood region.
3.   Urban area investment support for cities.
4.   Clean Energy for EU Islands initiative.
5.   Structural support for regions with high coal production and high carbon emissions.
6.   European initiative 'Young people for climate'.
7.   Smart building investment financing instrument.
8.   A compilation of EU regulations on the energy performance of public buildings.
9.   Investment in clean industrial technologies.
10.   Clean, competitive and network-based mobility.
None of the instruments indicated per se relate to agriculture, retention, flood or drought prevention. A group of intermediate measures reducing CO_2 emissions as well as reducing the carbon footprint are indicated. Long-term strategy to 2050, accessed: 10.08.2020, https://ec.europa.eu/clima/policies/strategies/2050_pl#tab-0-1.

was taken. Drought protection in water management planning – a methodology for dealing with, envisages drought prevention and mitigation. It assumes, among other things: the development of rules for the control of water retention facilities in such a way that these resources can be used in times of drought, to increase water retention in agricultural areas, to create awareness among farmers of the possibility of creating water retention in agricultural areas, and to implement investment measures for the creation/enhancement of artificial retention. It was indicated that the programme would involve a combination of all available water retention methods: large-scale retention, small-scale retention, artificial retention, natural retention and melioration. The appendix containing the adopted programme assumptions includes a list of 94 investments with an assumed value of approximately PLN 10 billion, which are due to be implemented by 2027.[9] The expected effects of the programme will be:

- "an increase in the volume of water retained;
- an increase in the capacity of small retention facilities;
- mitigation of the effects of drought with particular focus on rural areas
- and woodland areas;
- a reduction in flood risk, including that associated with so-called flash floods in urbanised areas;
- restoration or improvement of conditions for the use of water for electricity generation;
- an increase in the contribution of local and regional water retention projects;
- an increase in public awareness of the problem of dwindling water resources and the need for water retention;
- an improved environment for the agricultural exploitation of water;
- enhancement of ecosystems created or maintained through the use of water retention;
- an improvement in the class and stability of navigational conditions on inland waterways;
- an improvement of the landscape value of water-related areas.

Sources of funding:
- European funds for the 2021–2027 financial period;
- loans or credits granted by international financial institutions: World Bank, Council of Europe Development Bank, European Investment Bank,
- state budget,
- local authority budgets,
- budgets of other entities (e.g., Polish State Waterways, State Forests, National Fund for Environmental Protection and Water Management);
- public-private partnerships".

---

9   Ministerstwo Gospodarki Morskiej i Żeglugi Śródlądowej, accessed 11.08.2020, https://www.gov
    .pl/web/gospodarkamorska/rzad-przyjal-zalozenia-do-programu-rozwoju-retencji.

The range of investments included in the Assumptions for the Water Scarcity Programme ... is controversial in terms of their targeting. These are primarily related to the potential for exploiting water in large bodies of water. The solutions adopted as a benchmark correlate primarily with energy supply requirements (cooling of power generation plants and CHP units) and as a result, advance environmental depletion and biodegradation. *"In general: flooding and drought – these are climatic phenomena that can be adapted to by reducing the risk of their occurrence or can be mitigated by protecting the areas affected and adapting the standards and procedures of this protection to progressive climate change. However, since there is a shortage of water in bodies managed by the Ministry of Marine Economy and Inland Navigation (GMiŻŚ) due to defective management (cut and deepened river beds, dredged drainage ditches, lack of irrigation and dedicated retention, poor condition and ecological potential of the waters), what chance is there that even greater water deficits will not arise once new land is occupied and developed? Why is it that, to date, the land allocated for water under the management of the Ministry of National Water Management (KZGW) is mostly abandoned land, devoid of original habitat and overgrown with invasive alien species?*[10] Importantly, in the indicated investment approach, there is

---

10   3. *On page 8, the assumptions indicate that: "Among the 13 specific objectives of flood risk management, the initial flood risk management plans include the maintenance and enhancement of the existing retention capacity of the catchment areas in the water region and the reduction of the existing flood risk (implemented, among other things, through the construction of water retention facilities and the construction or restoration of drainage systems). In total, more than 1550 2 activities have been planned under the Flood Risk Management Plan (PZRP) amounting to a total of 11.6 billion PLN. Many of these investments serve to increase water retention." That is, the assumptions envisage a further deepening of the water deficit! The traditional facilities (dam storage reservoirs) built by the Ministry/KZGW collect water for their own purposes: mainly for the production of electricity, with only a small flood reserve, and instead of polders, so-called dry retention reservoirs are built, which are sometimes transformed into ordinary dam reservoirs a few years after construction (...). It is also important to remember that: – it is not possible to use the water from dam reservoirs for any purpose other than collecting it and possibly controlling the amount flowing through the reservoir. – the surface area of the reservoir does not contribute to any retention, as in Poland the annual evaporation of water from the water table usually equals the annual precipitation; this also applies to all "water reclaimed" gravel pits, sand pits and other incidental open water bodies. The same sites preserved as forests, fields, meadows or, better still, wetlands or marshes, would constitute the best means of retention to protect adjacent areas from drought – in addition to the contents of dam reservoirs, retention in a catchment area is considered to include the capacity of navigational facilities (waterways, barrages, stages and sluices) realised in river channels, which are known to reduce the flood reserve of regulated channels, reduce the retention capacity for flood waters of unregulated river channels and valleys and have no other economic significance (example: the planned navigation stage in Niepołomice). Another observation is the Włocławek dam reservoir and the permanent drought in Kujawy adjacent to this reservoir. The impossibility of allocating water retained in e.g., the Solina or Dobczyce dam reservoirs for agricultural irrigation purposes, or water in the Czorsztyńskie reservoir for making snow on the ski slopes in Bukowina or Białka, confirms that retention in dam reservoirs is pointless and without an addressee (stakeholder) for this retention.* J. Jeleński, Uwagi do założeń do programu rozwoju retencji na lata 2021–2027 z perspektywą do roku 2030 opracowanego przez Ministerstwo Gospodarki

a major limitation in the possibility of water exploitation for agricultural purposes. This concerns the issue of water quality and the question of the potential rationality and technical feasibility of its transport to remote areas (e.g., agricultural areas, from central Poland to its north-eastern part, e.g., Białystok). The problem of supplying agriculture with water for production can even arise on a regional or local scale, so the philosophy of the model presented is that:

1.   We retain water at the point of precipitation/runoff and as close as possible to the point of use in terms of agricultural, environmental, economic or emergency management use.
2.   Water retention at the micro (micro catchment) scale translates into flood risk prevention at the macro scale (river basins of the large Polish rivers Oder, Vistula).

2   Economic impact of drought on social stakeholders

Poland has an area of 312,679 km. More than 60% of the country's total area, i.e., 1,468,200 ha, was under cultivation in 2020.[11] Of this figure, 3.3 % is represented by organic (ecological) agriculture which is equally vulnerable to water scarcity.[12] The value of the gross domestic product generated in 2020 was 12,657,190,000 PLN. The number of people working in agriculture in 2019 was 1624.2 people, or 9.2 % of the population, against an EU average of 3.9 % of the population.[13]

The economic impact of drought on Poland has two dimensions – direct and indirect. Both are correlated with each other impacting on the economic health of farmers/businesses (agriculture), the environment (biodiversity), and society. The issue concerns the environment and the scale of the causes and effects of its changes in a specific spatial area, the magnitude of their potential impacts, their duration, their reversibility, the measurability of the accompanying factors and processes, as well as the degree of their complexity, connectivity, networkability.[14] These elements are interconnected with further elements such as: "resilient communities", "climate resilient development pathways", "resilient future" and finally

---

Morskiej i Żeglugi Śródlądowej, Stowarzyszenie Ab Ovo, Kraków, ul Chodkiewicza 14, accessed 11.08.2020, http://praworzeki.eko-unia.org.pl/imgturysta/files/2019-06-21%20Uwagi%20do%20Programu%20rozwoju%20retencji%202021%20-%202030.pdf.

11   *Rocznik Statystyczny Rolnictwa 2021*, accessed: 15.07.2022, https://stat.gov.pl/obszary-tematyczne/roczniki-statystyczne/roczniki-statystyczne/rocznik-statystyczny-rolnictwa-2021,6,15.html.

12   Eurostat, accessed 14.08.2020, https://ec.europa.eu/eurostat/cache/digpub/keyfigures/#.

13   *Rocznik Statystyczny Rolnictwa 2021*, 400.

14   Jon Barnett, Stephen Dovers, *Environmental security, sustainability and policy*, Pacifica review: peace, security & global change 13:2 (2001): 157–169, https://doi.org/10.1080/713604521.

"resilient planet".[15] The forms of the narrative are variable, what remains constant is the context of environmental protection, and of adaptation to climate change by humanity, individual societies, communities.

As the climate in Poland warms, there is a progressive decline in groundwater levels due to reduced rainfall and soil degradation. Together and individually, each of these sets a direct economic cost, not least in terms of a deficit in the water needed and available in a given area.[16] This is expressed in the decreasing incomes generated by land users as a result of lower land productivity.[17] This situation is closely linked to environmental degradation based on access to water. This occurs as a result of a change in the conditions of their functioning. These costs occur 'on the ground' and are felt by everyone, from the land user to the local community, its livelihood, its development. This has an indirect impact and varies in each region of Poland.[18]

The situation described could change. There is an opportunity to functionally link water retention to the state's emergency management system. We assume that this could have a significant impact:

– reducing the risk of natural disasters such as drought or flooding,
– reducing economic and environmental losses caused by drought or flooding, and:
– taking into account, in the determination of the gross domestic agricultural product by the Central Statistical Office, through direct, local water restriction costs based on biophysical measurements (e.g., metering systems set up in the construction of small and medium-sized retention facilities within microcatchments).
– economic research into the profitability of growing crops in specific areas subject to progressive degradation of the soil as a result of prolonged drought.
– implementation of new technologies related to water retention in domains such as:
  – sensors (for various applications),

---

15   Janpeter Schilling, Sarah Louise Nash, Tobias Ide, Jurgen Scheffran, Rebecca Froese, Pina von Prondzinski, *Resilience and environmental security: towards joint application in peacebuilding*, *Global change*, peace & security 29:2 (2017): 109–114, DOI: 10.1080/14781158.2017.1305347.

16   Ekologia.pl, accessed: 14.08.2020, https://www.ekologia.pl/wiedza/slowniki/leksykon-ekologii-i -ochrony-srodowiska/deficyt-wody.

17   "Widespread drought resulted in significantly lower yields for most crops in 2018. In particular, cereal yields fell sharply (–16.1%) to 26.8 million tonnes, despite larger areas being cultivated for almost all the main cereals. Lower yields were harvested for all major cereals: wheat (–15.8%), rye and winter (–18.5%), barley (–19.6%), oats and spring (–15.0%) and grain maize (–3.9%). Poland became the biggest producer of rye in the EU in 2018 and the second largest for oats". *Agriculture, forestry and fishery statistics 2019 edition*, European Union (2019): 186.

18   Joanne Linnerooth-Bayer, Anna Dubel, Jan Sendzimir, Stefan Hochrainer-Stigler, "*Challenges for mainstreaming climate change into EU flood and drought policy: water retention measures in the Warta River Basin.*"

- cyber-security,
- autonomous vehicles (ground, air, as well as underwater and on the surface),
- education to create new social competences.
- science to implement homegrown technical, technological, IT, cyber security infrastructure solutions.
- strengthening the organisational and strategic culture of various public and private entities.

When considering the economic determinants of water protection and retention in crisis management systems as an antidote to the persistent drought phenomenon, the role of state policies and strategies, including – as a derivative – social, economic, others – security, should be noted. These diverse functions have a common interpretation and a limitation in the imagination of farmers, politicians, economists, scientists, users, designers, and society.

Firstly, it is important to note the constant endeavour of state institutions and the European Union in the context of meeting the challenges of change in their environment as well as in themselves. In this case, the issue concerns:

- shaping the structure of the economy,
- shaping of the agricultural system and agricultural policy,
- the transformation of traditional industries (agriculture) and their modernisation,
- developing new forms of technology, techniques (software, sensors, equipment, machinery), their coupling to the emergency management system together with the water protection and retention subsystem.

This imperative to act should be at the core of the actions taken by the institutions of the state and the Union concerning space, the people living in it, and the natural environment.[19] It carries important cognitive implications and creates a paradigm of knowledge and ignorance (as important as the former, as it enables the verification of actions, and in them challenges, threats, and opportunities).

Secondly, it is important to draw attention to the continuing desire of state institutions to build and maintain competitive advantage. This is done by creating the right conditions for economic development to take place. In the case of water retention in the crisis management system, the strengths of the EU countries consist from: skilled workers, innovative farmers/agricultural entrepreneurs, entrepreneurial Europeans, relatively high levels of technical and economic education, and technologies easily adopted by farmers/entrepreneurs. Constraints include an underdeveloped digital infrastructure (currently, the Internet of Things is a terra incognito),[20] a lack of knowledge among policy-makers, and economic elites, which

---

19   Josh Watkins, "Spatial imaginaries research in geography: synergies, tensions, and new directions," Geography compass 9, no. 9 (2015): 508–522, https://doi.org/10.1111/gec3.12228.

20   The Internet of Things (IoT) implies the networking of almost all types of devices. Related to this notion is the vision of a future world in which digital and physical devices and everyday

constitutes an element of situational awareness (blessed are the ignorant, for theirs is the kingdom of ignorance), and a lack of a developed industrial base for the creation of autonomously moving vehicles (on the ground, in the air, and perhaps on water and underwater) fully integrated with digital technologies. This form of state activity encourages its systemic treatment through the lens of social structures. These structures will be different in each case depending on how and in what social environment the state implements its activities. With the right policy and strategy in place, the state creates a security space of its own, in which it undertakes activities aimed at achieving its security objectives, as well as their derivatives.

Thirdly, the continuous acquisition and targeted use of indigenous resources should be included and identified as a priority. The EU's environmental, agricultural and economic policies should evolve to support the technological competitiveness of specific industries, and in this case agriculture. Water retention and conservation in the state's crisis management system creates the conditions for cooperation between public institutions, research centres, businesses and representatives of society. The outcome can be the mobilisation of the economic resources of individual states, as well as the European Union, together with the skills, knowledge, abilities and predispositions of society including farmers/entrepreneurs. The role of communication and the exchange of information, which play a fundamental role in decision-making processes, should be recognised here. In addition, the basis for creating competitiveness through the use of natural, renewable resources (which can consistently provide useful outputs). Importantly, state support of a structural nature, despite the outlay, should produce ongoing returns in various forms, but primarily environmental and economic.

Fourthly, the creation of a competitive advantage for European farmers/entrepreneurs. The basic consideration is to establish the nature of the social and political relationship, and interaction. Water retention in the crisis management system translates into an increased degree of autonomy of action for farmers/businesses. It gives them the advantage of a resource – water – and a choice of ways to use it for development. What is more, it allows freedom to use financial instruments (e.g., equity, reimbursement of eligible costs of operations, area payments, de minimis aid, loans, leases, investment funds) to achieve higher returns, i.e., a higher return

---

objects are connected by an enabling infrastructure. The aim is to deliver a variety of new applications and services. The Internet of Things is the combination of two worlds as we know them, and over the coming years, we will see a fundamental shift in the way we use and interact with both the digital device world around us and the physical world. The concept of the Internet of Things is based on three ideas: always (anytime), everywhere (anyplace), and with everything (anything). The definition of the Internet of Things is based on three pillars relating to the characteristics of smart objects: enable self-identification (everything is able to introduce itself), provide communication (everything can communicate) and interact (everything can interact with each other). A. Brachman, *Internet przedmiotów*, Raport Obserwatorium ICT, (Technopark, Gliwice 2013), 7, after: Andrzej Kobyliński, "Internet przedmiotów: szanse i zagrożenia," Ekonomiczne problemy usług, Zeszyty Naukowe Uniwersytetu Szczecińskiego 112 (2014): 102.

on invested capital. This undoubtedly creates advantages for farmers/entrepreneurs in terms of the quality of their produce and its price. An important element of this approach is the creation of information superiority. By linking water retention to environmental protection, the green economy creates an additional competitive advantage. Clearly, water retention in the crisis management system is intended to reduce environmental risks and ecological deficits. Its introduction implies sustainable development without environmental degradation. This creates an information advantage that can easily be used in media messaging, descriptions of agricultural products (specific trademark) that can also reach the consumer. In addition, this type of advantage can be used to shape consumer preferences more effectively than competitors in other countries or regions of the world at a given level of product quality and price.

Fifthly, the above-described elements can form the intellectual capital of Polish farmers / entrepreneurs, and representatives of public institutions (administration and science). The connections between knowledge, lessons learned, customer interaction formula, organisational capacities, forms of communication, and public-private cooperation are obvious. They will enable the acquisition of new qualifications, competencies, practical skills, relationships, connections, correlations creating human capital, adding value to the Polish economy, as well as intellectual and industrial values. In the case of the state, they will additionally form competences in the national security system that determine its form in time and space. Above all in terms of organisational culture, decision-making structures.[21]

Taking into account the technical and economic aspects of water retention and drought prevention in crisis management systems, there is a chance to create structural capital in the form of knowledge (know – how), patents, licenses, software, technical and technological thought. This can lead to increased innovation in small and medium-sized enterprises, strengthening their market position and increasing employment. This could be the case in biotechnology, information and communications, materials research, photonics, manufacturing technology, resource efficiency, medicine, climate protection and security. The effect of the aforementioned will be the advancement of the Polish economy through the creation of relational capital with economic actors in and around the agricultural sector, e.g., financial institutions, credit institutions. Furthermore, creating and strengthening the reputation of the Polish economy. In consequence, the creation of its tangible and intangible assets, innovation, enhancement of the quality of products and services. This element should translate into a competitive advantage for Polish farmers/entrepreneurs in the market and, overall, a competitive position for Poland in the global market.

Sixthly, water retention and drought prevention in the European Union's crisis management systems create opportunities for the development of specifically

---

21   Jarosław Gryz, Aneta Nowakowska – Krystman, Łukasz Boguszewski, *Kluczowe kompetencje systemu bezpieczeństwa narodowego*, (Wydawnictwo Difin, Warszawa 2017), 15–104.

TABLE 2.1   Internet of Things about defining patterns of behaviour, building a product

Ways to behave in the Internet of Things

		Traditional set of values	New set of values
Making values	Individual needs	Solutions for existing needs, lifestyle, reactive formula	Actual, immediate provision of anticipated needs
	Offer	Aging of the product	Continuous product improvement
	Importance of data	Single datasets are used for future products	Instant convergence of information from various sources
Capturing values	Way to profit	Selling of another product or item	Possibilities of duplicating income
	Check points	Potential advantages of the goods (ownership of the logo, etc.)	Personalization and context (centric network effects)
	Development abilities	Use of key competences, existing resources and procedures	Adaptation of income duplication in different systems

SOURCE: COMPILED FROM SUBJECT LITERATURE

profiled fields of study in universities, economics, agriculture, or homeland security. There is an important requirement for Poland and the European Union to adapt to the fourth and, above all, related fifth industrial revolution based on the Internet of Things (individualism – technical cooperation – knowledge – skills and competences), (table 2.1). This is the only place where the 'farm-to-table' demand of ordering produce directly from the farmer using visuals (I'm ordering this apple from this tree), delivered by drones 'to your doorstep', can be seen. The quality of this education will determine the economic position of the country in the future. Three elements come to the fore here. Firstly, the individualism of Poles that can be easily related to individual usage of the Internet of Things – farmers / entrepreneurs. Secondly economic. In this respect, a system of tax exemptions should be created in Poland to allow farmers/entrepreneurs to invest in infrastructure and research into water retention and not only for it, but for agriculture as a whole and its associated industries. Thirdly education with an indication of the benefits of the Internet of Things.

Seventhly, the inclusion of water retention in the crisis management system opens the field to the development of existing areas of science: hydrology, land reclamation, biology, agricultural economics, technology, hydraulic engineering, computer science, administration, management, internal security, national and international security, and others. Additionally, the emergence of new research areas and fields, e.g., the economics of water retention.

Eighthly, the social structures involved in the processes of water retention and conservation in the EU crisis management system will serve to transpose social roles and relations. They make it possible to create a new model of political decision-making which will be of particular importance not only for local communities but also for community organisations. Embracing the interpretation of organisational and management sciences, and through this, a new shape will be given to functional management (through objectives, results of actions) and institutional management (mission and vision of the state, basis and form of expected changes). Both of these are fundamental to the formulation and implementation of security policies and strategies implemented across time and space. It is usually an open question to what extent the adopted format of strategic security management is realistic, and able to combine the mission and vision of the state, and in this particular case crisis management.

Ninthly, water retention and drought prevention in the security management systems of Poland and the European Union form the basis of Polish and EU economic policy not only in national but also in international terms. The catalogue of guidelines related to water retention, and drought prevention, corresponds to the range of challenges faced by the state, the European Union and their institutions. They are as follows:

- traditions, knowledge and skills that require adaptation to the changes formulated in this work.
- norms in the legal system defining the role and nature of water retention in the crisis management system and state security more broadly.
- consistency in the implementation by public institutions of the identified tasks within the framework of strategic security management.
- the obligatory need to verify the achievements and failures associated with the implementation of the concept by public institutions.

It may seem that this catalogue of challenges has nothing or very little to do with the economic impact of water retention and drought prevention on social actors in Poland or in the European Union, but this is not the case. Indeed, the nature of the challenges is not in an economic character, but in a political one, which defines the former. Water retention, and drought prevention, is a projection of political economy, or lack thereof. This situational aspect determines its relative rank, in theory and in practice.

3 Policy challenges associated with the economic impact of drought

The policy challenges for Poland and the European Union related to the economic impacts of drought are complex. They result from the way the state is organised, its administrative division, the configuration of public institutions, the participation of citizens in social and political life, and finally social and political groups, and elites. Each of these elements determines the activities of the others.[22] This truism also applies to the projection of Poland's economic development since the political transformation of 1989 and, since 2004, within the European Union. This development has reconfigured the way the state operates by bringing Polish economic actors into international production chains. This has also applied to agriculture as a branch of the Polish economy, which has been part of the European Union's Common Agricultural Policy since 2004, financed until 2007 by the European Agricultural Guidance and Guarantee Fund and now by the European Agricultural Fund for Rural Development (financial support for rural agricultural policy measures). In addition, the European Agricultural Guarantee Fund (financing of measures under the so-called first pillar of the CAP, i.e., area payments)[23] and the state budget in the supplementary part. During this period, the following were implemented in Poland as part of assistance from the aforementioned funds:

- Rural Development Programme 2014–2020 (was extended for period 2021–2022),[24]
- Rural Development Programme 2007–2013,[25]
- Sectoral Operational Programme (SOP) "Restructuring and modernisation of the food sector and rural development 2004–2006" implemented under the Rural Development Plan (RDP) as part of the National Development Plan.[26]

---

22   Sune Wiingaard Stoustrup, The re-coding of rural development rationality: tracing EU governmentality and europeanisation at the local level, European planning studies (2021): 1–14, https://doi.org/10.1080/09654313.2021.2009776.

23   *Rozporządzenie Parlamentu Europejskiego i Rady (UE) nr 1306/2013 z dnia 17 grudnia 2013 w sprawie finansowania wspólnej polityki rolnej, zarządzania nią i monitorowania jej oraz uchylające rozporządzenia Rady (EWG) nr 352/78, (WE) nr 165/94, (WE) nr 2799/98, (WE) nr 814/2000, (WE) nr 1290/2005 i (WE) nr 485/2008, L 347/549–557.* "To ensure sustainable rural development, it is necessary to focus on a limited number of core priorities related to knowledge transfer and innovation in agriculture, forestry and rural areas, farm viability, competitiveness of all types of agriculture in all regions and promotion of innovative technologies on farms and sustainable forest management, organisation of food chains including the processing and marketing of agricultural products, animal welfare, risk management in agriculture, restoring, preserving and enhancing ecosystems related to agriculture and forestry, promoting resource efficiency and the transition to a low-carbon economy in the agricultural, food and forestry sectors, and fostering social inclusion, poverty reduction and promoting economic development in rural areas." *Rozporządzenie Parlamentu Europejskiego i Rady (UE) nr 1305/2013 z dnia 17 grudnia 2013 r. w sprawie wsparcia rozwoju obszarów wiejskich przez Europejski Fundusz Rolny na rzecz Rozwoju Obszarów Wiejskich (EFRROW) i uchylające rozporządzenie Rady (WE) nr 1698/2005, L 347/487.*

24   ARiMR, PROW 2014–2020, accessed 05.09.2020, https://www.arimr.gov.pl/pomoc-unijna/prow-2014-2020.html.

25   ARiMR, PROW 2007–2013, accessed 05.09.2020, https://www.arimr.gov.pl/programy-2002-2013/prow-2007-2013.html.

26   ARiMR, SPO Rolnictwo, accessed 05.09.2020, https://www.arimr.gov.pl/programy-2002-2013/spo-rolnictwo-2004-2006.html.

Concurrent programmes of the European Union's Common Fisheries Policy were implemented, including;
- Fisheries and Maritime Operational Programme 2014–2020 (ongoing)[27]
- Operational Program "Sustainable development of the fishing sector and coastal fishing areas 2007–2013",[28]
- Sectoral Operational Programme "Fisheries and fish processing 2004–2006".[29]

Poland has also benefited from support for the adaptation of the agri-food sector to the requirements set out in European Union legislation under the Special Accession Programme for Agricultural and Rural Development (SAPARD) programme, implemented since 2002, Special Accession Programme for Agriculture and Rural Development (SAPARD).[30] Thanks to these programmes, implemented by the Agency for the Restructuring and Modernisation of Agriculture, agriculture and fisheries in Poland have been given the tools to support their development over time.

As a consequence of including Polish agriculture within the European Union's agricultural policy and market system, the production process of goods and services has become part of the global market circulation. What is important is that the most processed ones, related to mechanisation, are the result of ideas and technologies, of design, produced in the most developed countries of the planet including the European Union. Poland's current development stems from a phenomenon in which Polish economic players are producing increasingly high-end goods and services yet are still largely duplicating the experience of their EU and non-EU partners, entities with which they cooperate and compete. Agriculture linked to water retention could become an opportunity and an incentive for them. This could take place in the state's crisis management system, or perhaps in any of its other systems – it is just a matter of imagination related to the configuration of activities within its policies and strategies. There is one prerequisite for this – the system must involve agriculture and make water retention usable not only at the scale of the individual producer/businessman, but also at the level of the local community, and the region. This factor alone will determine whether ideas and technologies related to water retention will become part of Poland's economic growth in many sectors. Its strengths, a skilled and ambitious workforce, entrepreneurship, forward-looking agriculture and its natural environment, predestine this.

---

27 ARiMR, Ryby 2014–2020, accessed 05.09.2020, https://www.arimr.gov.pl/pomoc-unijna/pro gram-rybactwo-i-morze-2014-2020.html.

28 ARiMR, PO Ryby 2007–2013, accessed 05.09.2020, https://www.arimr.gov.pl/programy-2002 -2013/program-operacyjny-ryby-2007-2013.html.

29 ARiMR, SPO Rybołówstwo i przetwórstwo ryb 2004–2006, accessed 05.09.2020, https://www .arimr.gov.pl/programy-2002-2013/spo-rybolowstwo-i-przetworstwo-ryb-2004-2006.html.

30 ARiMR, SAPARD, accessed 05.09.2020, https://www.arimr.gov.pl/programy-2002-2013/sapard .html.

Provisions were included in the Responsible Development Strategy that speak of the low innovativeness of the Polish economy, including barriers of a systemic nature, such the legal and institutional environment, which does not stimulate risky innovative activity, as well as insufficient coordination of activities and support instruments (at national and regional levels).[31] As a key to economic development, water retention in Poland's crisis management system addresses the need for redefining it. However, this requires breaking orthodox thought patterns and being assertive in our relations with the world, above all with the highly industrialised countries of our planet. Domains such as IT and cyber-security, energy technologies based on water, wind, solar (point source power for infrastructure in the microgrid area), equipment and machinery, can become enablers of Poland's economic development if this challenge is taken up. The cited document indicates that insufficient cooperation of institutions carrying out public tasks in the field of innovation, duplication of tasks and responsibilities and imperfect procedures for assessing the innovativeness of individual solutions are also a problem, with the knowledge management of external experts evaluating innovative projects requiring significant improvement.[32] The reason for this is "ineffective communication between science and business and inadequate cooperation in the area of innovative activity (which, author's note) results in an insufficient market potential of the conducted research, low interest among entrepreneurs in the research work of the world of science and its results and, consequently, a preference for implementation of ready-made solutions. In the context of eco-innovation, an additional barrier is the low awareness among manufacturers of the benefits associated with their implementation in terms of economies – lower costs of running a business."[33]

The current state of the Polish economy does not inspire optimism, 'every year 95 billion PLN flows out of Poland abroad as dividends, and two-thirds of Polish exports and half of industrial production are the work of companies based on foreign capital. Further, the pitfall of the mediocre product is the result of Polish companies not being keen on innovative activity. Polish R&D expenditure in relation to GDP is twice as low as the EU average. Only 13 per cent of the Polish SME sector innovates, while the EU average is 31 per cent. Undoubtedly, this particular pitfall is an offshoot of the former – companies on the Vistula have no special incentive to look for more profitable solutions when they still have a pool of low-cost workers readily available.'[34] Therefore, placing emphasis on the development of agriculture configured with ecology, and security, creates an opportunity to overcome the described state of affairs, if only partially, and in the long term – perhaps for this industry and

---

31   *Strategia na rzecz Odpowiedzialnego Rozwoju*, 87.

32   Ibid.

33   Ibid.

34   Piotr Wójcik, *Uciec z półperyferii – polityka przemysłowa PiS*, in: *Ekonomia polityczna „dobrej zmiany"*, ed. Michał Sutowski (Warszawa: Instytut Studiów Zaawansowanych, 2017), 161.

those operating around it – completely. Science – economics – development, this triumvirate has a chance to appear as mutually equivalent in the Polish crisis management system. In doing so, it can incorporate water retention as a second (compatible) pillar – alongside flood prevention – into its operating formula. This requires policymakers to have the courage and vision to develop the state using local public-private partnerships.[35]

By treating water retention in the crisis management system as an instrument for shaping the economic policy of the state, institutions of political power can create an opportunity for:
- a specific, targeted economic policy to change the structure of Polish production (in the configuration: strategic management – policy – strategy / science – techniques – technologies) to a more advanced one,
- support for the development of innovative technologies with applications at the interface between the environment, agriculture and security, which can help develop small and medium-sized enterprises. If they are successful, they should be protected from being taken over by foreign capital and thus impoverishing the Polish economy, innovation, and their contribution to the development and security of the state,
- support for the development of social innovation in strategic planning and management,
- crisis management incorporating water retention creates an opportunity for agricultural development with a high potential for sales growth whilst protecting the environment, the use of advanced technology (Internet of Things, Artificial Intelligence), advanced management processes, low resource consumption, and the requirement to undertake extensive cooperation between diverse actors.

The indicated scope of activities requires the government to coordinate the development of new industries correlated with water retention in the security management system of the state as well as the European Union. The indicated scope of activities requires the government to coordinate the development of new industries correlated with water retention in the security management system of the state as well as the European Union. Creating this interpretation requires:
- the reconfiguration of social structures within public-private-social partnerships, norms and laws that determine social behaviour including political power structures,
- to identify society's preferred values, needs, interests and objectives identified through the lens of development,

---

35   Vivien Lowndes, Helen Sullivan, "Like a horse and carriage or a fish on a bicycle: how well do local partnerships and public participation go together?," Local government studies 30, no. 1, (2004), 51–73, DOI:10.1080/0300393042000230920.

- redefining state security as embodied in political ideas, ideology, doctrines, in the programmes of political parties, groups in power, their representatives, and political culture.
- to define the exercise of power, the implementation of policies and strategies of the state, including its security.

The factor that binds the above-mentioned elements together is the state's mission and its vision, a projection of development over time in terms of providing optimal social, political, economic, other conditions.

A practical question-and-answer framework for policymakers can be used to verify the concept of water retention in Poland's crisis management system. These are as follows:

- What objectives do we want to achieve in this particular policy and why?
- What actions should be taken, what resources will be required for this and how much will they cost?
- What are the reasons for believing that the proposed policy actions will achieve the stated objectives?
- What is the potential, intended, and other consequences of this policy?
- What is the degree of effectiveness of this particular policy?
- What will be the cost-effectiveness of this particular policy – will a balance be maintained between the expenses incurred and the objectives achieved?
- What is the risk that this particular policy will fail or even worsen the current situation?
- What are the risks involved if we achieve our stated objectives that the policy will harm other objectives in other areas and – perhaps – do more harm than good?
- If we consider all of the above, will this particular form of policy ensure that objectives are satisfactorily achieved at an acceptable cost?
- How does this particular policy compare to other policy options which seek to achieve the same objectives with different forms of action and costs?
- Will another policy be equally effective and less expensive, or will it be the same and achieve more?
- If among the several options selected there is no clear choice between them, how should they be evaluated?
- On balance, which option makes the most sense?[36]

Moreover, what assessment will future generations make of current state policy if the proposed direction is abandoned or even rejected for reasons of short-term political interests? The above list makes it possible to identify the most relevant elements, determining the approach to thinking and acting in the field of state security. Water retention is essentially about this. It addresses the most fundamental issue – the country's food security.

---

36   Richard L. Kugler, *Policy analysis in national security affaires: new methods for new era* (Washington D.C.: National Defence University Press, 2006), 13.

Here, it is important to recognise the range of practical undertakings related to the implementation of the concept of water retention in Poland's crisis management system. It concerns the implementation of the idea and the political oversight resulting from the concept. In doing so, they provide an interpretation of the desired functioning of the political institutions of the state. The following should be indicated here:

1. *Strategic planning.* This concerns the formulation of concepts for strategic preparations covering agriculture, the environment, finance, and forms of political, social, economic involvement, as well as for security institutions.

2. *Formulating action scenarios.* For policymakers, scenarios play an important role in determining rational courses of action in their decision-making processes. Essentially, they relate to the value system of decision-makers, the nature of the influence that is exerted on them, and the instruments for specific actions envisaged in state policy and strategy. Consequently, these scenarios are given a pragmatic interpretation by defining preferred courses of action. This is achieved by:

   a. Identifying the guiding assumptions of the changes to be implemented.

   b. Developing, based on the guiding assumptions, future studies in which the identified changes will be used.

   c. Verifying the implications of the assumptions made in the process of their implementation by analysing specific case studies, their comparisons, and the conclusions reached.

3. *Identifying the phases of implementation in time and space.*

4. Preparing public administration personnel (in the case of Poland, the role of the National School of Public Administration is important, and also – by conducting Higher Defence Courses – the University of Military Studies).

5. *Controlling the way things are done.* A spectrum of options for dealing with decision-making conditions is an open directory of activities. Policy makers should make their choice guided by the principles of expediency, rationality, cost-effectiveness and the expected end result.

6. *Monitoring events, phenomena, processes*, deriving appropriate judgements in relation to them in the context of identifying challenges, threats and opportunities associated with the actions taken and implemented.

The fact that various types of solutions feature in water retention and drought prevention within the state management system implies, first and foremost, the need for a new state economic policy in this area. This requires the diagnostics indicated above related to the development of agriculture, the sectors working in its support, as well as the public administration model. The measures indicated should be reflected in the state's policies and strategies and additionally in normative changes to both laws, regulations and lower order acts, e.g., local zoning plans.

When considering the forecast of economic development in Poland, as well as in the European Union, based on water retention and drought prevention, the most appropriate, comprehensive approach to this issue is capacity-based planning. This enables the use of specific, time-appropriate powers and resources. This method is used to look for capabilities that will provide freedom of action in terms of a wide range of security challenges and threats, creating opportunities in terms of security provision. Using this method for projecting social and economic development, it is possible to:

1. Define a specific environment and within it a course of action (using modelling methods and collections). This is particularly relevant for the preparation of farmers and other actors to undertake activities, localising micro-retention, and preparing administrative staff (from the municipal level to the central institutions of the state).

2. Identify future scenarios resulting from the introduction of water retention into the state's crisis management system. Draw on scenarios, the most preferred, most and least realistic, as a reference point for policies and strategies.

3. Identify non-linear determinants that may influence the initial assumptions made (e.g., different approaches to the retention concept in different provinces, and communities).

4. Define courses of action by identifying:
    a. Risk assessments, sources of threats to the implementation of the water retention concept in the crisis management system.
    b. An evaluation of ways to address the various risks identified with water retention and their associated risks.
    c. The nature of the impact of the above for ways of implementing the concept of water retention in the crisis management system.
    d. Ways to mitigate the impact of the various risks identified with water retention in the crisis management system, and their associated risks.

The use of the above-mentioned methods – capacity-based planning, modeling, sets, scenarios – will allow you to create the basis for making long-term decisions expressed in the form of a strategy and verify it over time. The above issues are evident in the definition of the state's security situation, both the current one and the desired future one.

4   Conclusions

The creation of conceptual foundations for the activities of the state, the institutions of the European Union, and the recognition of factors, processes, phenomena, trends concern society as a whole, representatives of the centres of power, as well as the power elite. The correct identification of the internal and external determinants of state policy, the perception of the degree of organisation of the social space, and its nature, determines the courses of action related to the creation of desirable political, social, economic, bureaucratic, intellectual attitudes, and the application

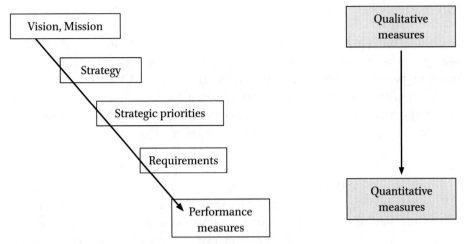

FIGURE 2.1 Strategy for measurable metrics
MODIFIED TROUGH TRANSLATION, ORIGINAL SOURCE: ANETA NOWAKOWSKA –
KRYSTMAN, *STRATEGICZNA KARTA WYNIKÓW JAKO NARZĘDZIE STRATEGII*,
IN *STRATEGIA BEZPIECZEŃSTWA NARODOWEGO POLSKI*, ED. JAROSŁAW GRYZ
(WARSZAWA: PWN WYDAWNICTWA NAUKOWE, 2013), 116

and adaptation of norms. Water retention in Poland's crisis management system makes it possible to introduce specific paradigms for perceiving reality creating a state of individual awareness and, consequently, collective social behaviour. This takes place through governmental organisations and institutions, which are part of the administration, and non-governmental organisations and institutions involved in political communication processes.

In the case of water retention and drought prevention, the requirement to have an adequate policy and strategy is self-evident. Working on them, as well as the accompanying debate, has the potential to create an extremely beneficial formula for the expression of defined views, their confrontation and, consequently, the creation of a coherent interpretation of the concept of action. Thanks to them, as well as community organisations of various types, it becomes possible to identify society-wide, long-term policy goals that have the potential to be integrated into a strategy for ensuring the existence and development of the state. Through this interpretation, the creation of a vision of the state, its security, and its subsequent realisation takes place.

Taking into account the fact that water retention and drought prevention can create a development trend for the Polish state and the European Union, it is worth demonstrating how it can be validated. It can be used to identify actions that should take place in the political institutions of the state. The rationale presented in this chapter can be captured in two forms – a vision and a development mission from which specific, operationalised goals are derived that make up the strategic concept of operations. These objectives can be framed in the form of an 'objectives tree' and their formulation included in the strategic tasks. In public administration, they may be assigned to specific posts, cells or individuals. The effect of this approach is to evaluate the effectiveness of the measures taken, to correct them or adapt them to new conditions that did not exist before. The tool for this will be monitoring, based on qualitative and quantitative measures (figure 2.1).[37] In the case of water retention in the state's crisis management system, things are straightforward in that performance measures relate to public entities, public-private partnerships (Partnerships for Water ...) and the actions of other social actors should they be involved. The processes and objectives are the basis for the supervision and derived evaluations as to the effectiveness of the projects implemented.

Processes These include institutions operating at the municipal, district, provincial and statewide levels. Their horizontal nature supported by information systems allows for basic who, when, how many, how, where information to be obtained. The issue relates to the relationship between vision, mission and strategy (i.e., the creation of a vision and strategic planning based on capabilities, the identification of priorities and within them goals, derived tasks, requirements for resource allocation, management and the creation of human capital). Against this background, the monitoring and evaluation of management control activities is demonstrable. This is because it creates awareness in the form of evaluations. Firstly, the conformity of the facts with the assumptions and requirements in the identified processes. Secondly, on the basis of the information collected and analysed, in terms of the activities of the organisations.[38]

The verification of water retention processes in the crisis management system can be done by means of a strategic scorecard (figure 2.2). Aneta Nowakowska-Krystman assumes that each of adopted perspectives has a set of measurable, sustainable, long-term and short-term goals. When used for verification, financial and non-financial yardsticks, operational performance indicators, and anticipatory

---

37   Aneta Nowakowska – Krystman, *Strategiczna karta wyników jako narzędzie strategii*, in *Strategia Bezpieczeństwa Narodowego Polski*, ed. Jarosław Gryz (Warszawa: PWN Wydawnictwa Naukowe, 2013).

38   In the case of Poland, the management control standards are contained in five groups corresponding to individual elements of management control. These are: a) Internal environment, b) Objectives and risk management, c) Control mechanisms, d) Information and communication, e) Monitoring and evaluation. In the case of public institutions, the responsible parties are the minister, for ensuring an adequate, effective and efficient management control system in the ministry (as head of the unit) and in the government department (as minister in charge of the department). The head of the municipality (mayor, city president), district starost, provincial marshal, who are respectively responsible for ensuring an adequate, effective and efficient management control system in the municipality (city office), district or provincial marshal's office, or in a local government unit. Communication No. 23 of the Minister of Finance dated 16 December 2009 on management control standards for the public finance sector. Annex to communication No. 23 of the Minister of Finance dated 16 December 2009, (item 84), 2.

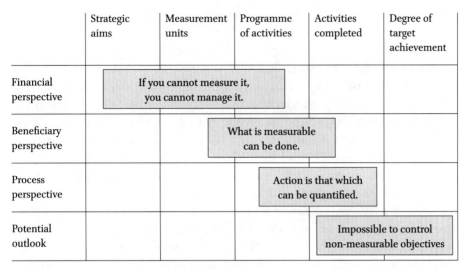

	Strategic aims	Measurement units	Programme of activities	Activities completed	Degree of target achievement
Financial perspective	If you cannot measure it, you cannot manage it.				
Beneficiary perspective		What is measurable can be done.			
Process perspective			Action is that which can be quantified.		
Potential outlook				Impossible to control non-measurable objectives	

FIGURE 2.2   The idea of measurable objectives in BSC
MODIFIED TROUGH TRANSLATION, ORIGINAL SOURCE: ANETA
NOWAKOWSKA – KRYSTMAN, *STRATEGICZNA KARTA WYNIKÓW JAKO NARZĘDZIE
STRATEGII*, IN *STRATEGIA BEZPIECZEŃSTWA NARODOWEGO POLSKI*, ED.
JAROSŁAW GRYZ (WARSZAWA: PWN WYDAWNICTWA NAUKOWE, 2013), 118

indicators can be used to establish the external and internal effectiveness of the strategy pursued, in terms of its practical implementation. This approach on the indicated perspectives prevents focusing only on a single measure. These have all been selected from a wide range of indicators and each is equally important to the management process influencing the implementation of the strategy. As a result, each of these perspectives assumes the existence of a set of measurable, sustainable, long-term and short-term objectives indicating financial and non-financial measures, indicators of operational performance, anticipatory indicators, external and internal effectiveness as a source of verification of the degree to which the objectives set have been achieved.[39]

Concluding the discussion of the political challenges related to the economic impact of water retention and drought prevention the links that define them are evident. Most significant is the relationship between policy and strategy of the state and that of the European Union. The nature of this relationship determines the implementation of the former and the effectiveness of the latter. The ability to ensure the security of the State and the European Union, shaping and giving the desired character is a fundamental challenge in defining the most appropriate strategy for actions identified for water retention and drought prevention. The correct

---

39   Aneta Nowakowska – Krystman, *Strategiczna karta wyników jako narzędzie strategii,* in
Strategia Bezpieczeństwa Narodowego Polski, ed. Jarosław Gryz (Warszawa: PWN Wydawnictwa
Naukowe, 2013), 118.

identification of factors common to the state and the European Union as well as their surroundings is a prerequisite for the success of social, political, economic and other security measures. What is more, this is a specific kind of progress that can become a development trend for Poland. The preparations made, the accompanying undertakings and their evaluations depend on: How the development policy will be pursued? How a particular form of strategy will be used to further enhance it? What, consequently, will be the results of both observable over time? What challenges, threats and opportunities will the European Union Member State be adapted to in the future? The end result of these preparations will be a real capacity of the state to ensure its security and furthermore the creation of social, political, economic and other realities.

The listed issues in the chapter are not the only ones that can be flagged. They formulate only an outline of the problems of this field of knowledge and require further research and more in-depth reflection. Moreover, verification of knowledge in relation to their utilitarian nature, as well as the practice of operation of institutions implementing policies and strategies for the development of the state and the European Union as a whole. It is primarily concerned with the instruments used to verify whether it is achieving its objectives, and whether the methods, techniques and means used to achieve this are effective and to what extent. The issue of implementation and realisation of the objectives of a state, or an organisation such as the European Union, for their security is closely related to political decision-making aimed at maintaining the ability to function effectively in an ever-changing environment, be it natural, social, economic, security or other.

CHAPTER 3

The LEADER method in the planning of water protection strategies

The water conservation strategy aims to identify the optimum formula for the selection of measures, and the means of achieving them over time. This approach is reflected in the creation of community competences oriented towards the achievement of set goals: analysing needs, increasing efficiency, competitiveness, and developing people's capacities. The LEADER method "Liason Entrée Actions de Development de l'Economic Rurale" was used here, which is a bottom-up, partnership approach to rural development, implemented through a Local Action Group (LAG). In this method, the basic task is to develop a Local Development Strategy (LSR) by a locally focused rural community, implementing joint, innovative projects, combining human, natural, cultural, historical, and other resources.[1]

1 Applying the LEADER method in the design and implementation of local development strategies – a case study

In 1991, the LEADER programme, as one of the four instruments of the Common Agricultural Policy (CAP), empowered local communities in the countries of the European Union. Its objectives were: to pursue sustainable development, stabilise employment, support rural development, and increase competitiveness (by exploiting potential and promoting the application of new technologies in rural areas).[2] This programme is currently utilised to facilitate the cooperation of local partners for the benefit of local communities. Through the programme, local communities are encouraged to realise the potential of the rural area in which they operate. It also contributes to the development of rural areas with the help of existing instruments of the Member State and the European Union in support of sustainable local development. It should be emphasised that the LEADER programme has not used other, additional, new methods since its launch. Therefore, in the solutions proposed, the water conservation strategy is intended to complement the programme. A new method of community development combining previous experience with

---

1   Mark B. Lapping, Mark Scott, The evolution of rural planning in the global north, in The Routledge companion to rural planning a handbook for practice, ed. Mark Scott, Nick Gallent, Menelaos Gkartzios, (Oxon/New York, NY: Routledge, 2019), 28–45.
2   Anne Macken-Walsh, "Partnership and subsidiarity? A case-study of farmers' participation in Contemporary EU governance and rural development initiatives," Rural society 21 no. 1 (October 2011), 43–53, https://doi.org/10.5172/rsj.2011.21.1.43.

new objectives of social development, economic development, protection of natural resources, the fight against global warming and other, derived objectives. All the local cooperation programmes implementing the LEADER initiative have so far been based on a core of building on a number of elements, i.e., partnership, local, bottom-up and integrated approaches, innovation, collaboration, networking of local partnerships, local funding and governance.[3]

The general legal framework for the current EU water conservation programmes is set out in Council Regulation (EC) No. 1260/1999 dated 21 June 1999 laying down general provisions on the Structural Funds, known as the General Resolution. The detailed guidelines for the LEADER+ project are set out in the European Commission's Notice to Member States dated 14 April 2000, as well as in Council Regulation (EC) No 1257/1999 dated 17 May 1999 on the granting of aid for rural areas from the European Agricultural Guidance and Guarantee Fund. In accordance with these, each Member State of the European Union is allowed to implement one national or several local/regional programmes. Member States designated an authority that was responsible for the preparation of the project and the guidelines regarding the LAG competition. The developed guidelines were submitted to the European Commission for approval, and it was up to the Commission to secure funding for the project. After approval by the Commission, the body responsible for organising the national project prepared a so-called programme supplement, which was no longer subject to examination by the Commission.

The LEADER+ programme developed by the state in question clearly described the procedure for the management and flow of public funding. An example is the legal framework for the support of rural development by the European Agricultural Fund for Rural Development under the LEADER Axis 2007–2013. They were defined by a Council regulation (WE) nr 1698/2005 from September 20 year 2005 on support for rural development by the European Agricultural Fund for Rural Development. In relation to Article 91 of the aforementioned Regulation, the European Commission indicated detailed implementing rules in Regulation No 1974/2006 dated 15 December 2006 laying down specific rules for the application of Council Regulation (EC) No 1698/2005 on support for rural development through the European Agricultural Fund for Rural Development.

Currently, the design and implementation of rural development programmes is based on a number of EU documents, which include:

– Commission Regulation (EC) No 1320/2006 dated 5 September 2006 laying down rules for the transition to the rural development support scheme provided for by Council Regulation (EC) No 1698/2005.

---

3 Leo Granberg, Kjell Andersson, Imre Kovách, Introduction: LEADER as an experiment in grassroots democracy, in: *Evaluating the European approach to rural development: grass-roots experiences of the LEADER programme*, ed. Leo Granberg, Kjell Andersson, Imre Kovách (Surrey: Ashgate Publishing 2015), 1–12.

- Council Regulation (EC) No 1290/2005 dated 21 June 2005 on the financing of the common agricultural policy.
- Commission Regulation (EC) No 1975/2006 dated 7 December 2006 laying down detailed rules for the implementation of Council Regulation (EC) No 1698/2005, as regards implementation of control procedures as well as cross-compliance in respect of rural development support measures.

Each European Union Member State, on the basis of Community guidelines, develops a National Strategic Plan (NSP) in cooperation with the EC. This provides a reference basis for the preparation of rural development programmes funded by the European Agricultural Fund for Rural Development, through which the plan is implemented. Each Member State can prepare and submit to the EC one national programme or alternatively regional programmes. Within the programme, priority axes can be distinguished through which rural development measures are to be implemented. These axes are improving the competitiveness of the agriculture and forestry sectors; improving the environment and the countryside; enhancing the quality of life in rural areas and diversification of the rural economy.

LAGs are selected by the competent authority designated by the Member State within two years of the approval of the rural development programme by the European Commission. The criteria used to select partnerships within a programme must be the same. Furthermore, the criteria within a programme when selecting a particular programme are also identical. These concern: the LAG partnership (representativeness, composition of the group, likelihood of implementation, administrative capacity); the local strategy (local character, dissemination of the idea, possibility of involvement and cooperation of local partners, links with other projects, coherence of the programme with regard to geographic, economic and social aspects, meeting the needs of the area, population served by the programme). The funds set aside under the Leader rural development programmes can be used to implement three measures, i.e.: 1 Implementing local development strategies. 2. Implementation of cooperative projects. 3. The functioning of Local Action Groups including skills acquisition, activation and LAG running costs.[4]

1.1   The Local Action Group in ensuring water resource protection

Local Action Group 'LAG' has been defined in the EU legal framework as a type of territorial partnership. The legally defined area of operation of LAGs is rural areas.[5] It brings together representatives of local partners from the public, private and

---

4   Roberta Sisto, Antonio Lopolito, Mathijs van Vliet, "Stakeholder participation in planning rural development strategies: using back casting to support Local Action Groups in complying with CLLD requirements," Land use policy 70 (2018), 442–450, https://doi.org/10.1016/j.landusepol.2017.11.022.

5   John Bachtler, Colin Wren, "Evaluation of European union cohesion policy: research questions and policy challenges," Regional Studies 40 no. 2 (2006), 143–153, https://doi.org/10.1080/00343400600600454.

non-governmental sectors to create a specific form of 'public-private-social' part-nership, activating the communities in which it exists.[6] In the case of European Union Member States, the legal basis for the operation of the LAG is EU legisla-tion and that of the body concerned. In Poland, these are the regulations governing associations and support for rural development.

In Poland, a member of a LAG may be a municipality, an educational establish-ment, a cultural institution, a parish, organisations and associations operating in a given area, including the Voluntary Fire Brigade, housewives' circles, companies, cooperatives, etc., and in association with ordinary residents. From 2004 to 2006, under the Sectoral Operational Programme 'Restructuring and Modernisation of the Food Sector and Rural Development' and from 2007 to 2013, a 'foundation' was allowed as a legal form of LAG. The solutions adopted by law make it possible to implement water conservation within the framework of public-private and social partnerships that can be established in a given area.

The leading role in the composition of the LAG is played by the local author-ity/rural, urban-rural or municipal municipality, which, when joining the LAG, defines within its boundaries its area of action and the number of inhabitants cov-ered by the initiative. Hence, the territorial area of a Local Action Group is defined by the boundaries of its member municipalities. This is important for determining the amount of funding a LAG can apply for under the LSR (the rate is calculated per capita). The area covered by the LAG and thus the implementation of the LSR is coherent and has a population of between 30,000 and 150,000. In the case of Poland, following a tender procedure, the provincial government selects the pre-pared LDS development strategies and signs an agreement with the LAG specifying the limits and rules for the use of funds under the Rural Development Programme. The governance of the LAG is carried out by the general meetings of the members, a board of directors and an internal control body (the Audit Committee). An addi-tional LAG body is the "Council", whose primary task, as defined in the strategy, is the selection of operations to be implemented under the LDS. Once the Local Action Group has prepared the Local Development Strategy and signed the agree-ment with the Provincial Government, it proceeds to implement the objectives set out in the strategy in accordance with the adopted areas and scope of operations. This is done by preparing potential beneficiaries to apply for the activities provided for in the LSR, launching competitions for the implementation of operations, con-ducting calls for proposals, and evaluations. In addition, the LAG benefits from sup-port for staff preparation, promotional activities, running costs, office functioning, activation and cooperation projects.

When considering the use of the LAG method of action resulting from the LEADER programme in a local water conservation strategy, it is important to iden-tify the characteristics that determine its utility. Firstly, bottom-up, the participation

---

6   Kaisu Kumpulainen, "The discursive construction of an active rural community," Community development journal, 52 no. 4 (2017), 611–627, DOI:10.1093/cdj/bsw009.

of the local community in the creation and implementation of the LDS, the locally led development strategy. Secondly, territoriality, the LDS is prepared for a socially coherent area delimited by the borders of the municipalities. Thirdly, integrating, connecting and cooperating with different local interest groups. Fourthly, partnership, the local action group as a local partnership integrates different actors from the public, social and economic sectors. Fifthly, innovation on a local scale, adapting modern technological, educational and social solutions. Sixthly, decentralisation of management and funding (in Poland, the LAG takes over the tasks of the provincial government in defining priorities, carrying out calls for proposals, selecting the winners of competitions, and preparing the ranking list of entities eligible for support) This aspect is particularly important in defining all these elements in the process of implementing, controlling and supervising the implementation of the water conservation strategy in keeping with local sustainable development. Seventhly, networking and cooperation including exchange of experience and dissemination of good practices within the framework of the NEB – National Rural Development Network and European networks.

1.2   *Sources of water resource conservation funding for Local Action Groups*
The source of funding for LEADER+ is the Member State budget and the Guidance Section of the European Agricultural Guidance and Guarantee Fund (EAGGF). The basic criterion for obtaining funding for rural development has so far been the evaluation of procedures and the way in which LAGs are selected which realistically ensure competition between partners. In addition, focusing funding on the best strategies that respond to unique environments and local needs.

Up to now, under the LEADER+ programme, funding has been available to countries that met certain cumulative criteria. These were and are:
– the existence of regional partnership-based groups (LAGs) that were able to apply for funding under this programme.
– elaboration of integrated local area development strategies by LAGs.
– the representativeness of the LAG, stemming from the participation of representatives drawn from the social and economic sectors.
– proof that the LAG is developing and implementing ideas and that partners are fulfilling the tasks assigned to them.
– partnerships in which the partners come from the area covered by the strategy.
– an area in which the strategy is to be implemented and which has adequate resources to implement it.
– a population of the area concerned is between 10,000 and 100,000 inhabitants.
– a population density of no more than 120 inhabitants per km2.
– a strategy that must be complementary to the programmes already in place in the area.

To date, the activities in LEADER+ have been focused within three areas. The first involved supporting pilot local development strategies that were based on the cooperation of local partners. The second area concerned inter-regional and international cooperation. The third area focused on rural networking, and it did not necessarily have to relate to areas participating in the LEADER+ programme.

1.3   *Implementation of the LEADER I, LEADER II, LEADER+ approaches –* *adopting the concept of water conservation action*

The first editions of the LEADER programme, sought to stimulate the development of rural areas of the European Community – the European Union most affected by the economic downturn (unemployment, population decline, restructuring of farms, etc.). Successive editions have pursued sustainable rural development by supporting and implementing integrated local development strategies, making use of the cooperation of local public, social and economic partners.

The LEADER I programme was implemented between 1991 and 1993. It initiated an innovative approach to rural development policy based on three pillars: territoriality, integration, and partnership. It also created opportunities to promote the involvement of local communities and to increase the diversity of rural areas according to their inherent characteristics. In subsequent years, 1994–1999, LEADER II programme encompassed the extension of projects created in the earlier programme. The main focus was on the implementation of innovative activities and the participation of groups involved in the development of rural areas. This programme reached approximately 50% of the rural areas of the then European Union.

After 2000, the LEADER+ programme was introduced. In doing so, the work initiated in previous editions was continued. The long-term aim of the project was to mobilise local communities to work together in realising the potential of rural areas. The tasks prompted them to work together and to develop and implement innovative ideas for the development of their local rural localities. This was intended to lead to a sustained increase in the level of development, as well as to an enhancement of their competitiveness. A necessary condition for the implementation of LEADER+ was the cooperation of various partners, both local and other national actors, including those from outside rural areas.

After three editions of the LEADER project, several features can be identified that are relevant to the issue of water conservation. These are:
– the activation and cooperation of local partners.
– the promotion of an integrated, local approach to the development of an area; and
– encouraging the development of innovative solutions.
– sharing experiences.
– supporting locally relevant small businesses.

Each of the identified features aligns with the concept of implementing local water partnerships within individual EU Member States. It pursues objectives that focus on the local community, its social and economic activation in support of an innovative approach to development.

1.4 Legal framework for the functioning of "The Land Flowing with Milk" LAG

In the case of Poland, LAGs have legal status and the ability to incur civil-law liabilities. During the 2004–2006 and 2007–2013 programme periods, they were registered in the National Court Register as associations (less frequently as foundations or associations of associations). For the years 2014–2020, the legal form of partnerships brought together under the Local Action Group was a 'special' association with legal identity Experience from previous periods of implementation of the Leader initiative under the SOP (Sectoral Operational Programme) "Agriculture" and RDP 2007–2013 led to the definition of a uniform organisational and legal structure for LAGs. It was introduced into the Polish legal system by the provisions of the Act dated 7 March 2007 on supporting rural development through the European Agricultural Fund for Rural Development (Journal of Laws 2013, item 173 j.t.). Additionally, it is also provided for in the Act on supporting the sustainable development of the fisheries sector with the participation of the European Fisheries Fund (Journal of Laws 2009, No. 72, item 619, as amended). The legal solutions adopted earlier have been preserved in the Local Development Act.

In Poland, in the 'special' associations established under the LAG formula, ordinary members may be natural persons, legal persons (foundations, associations, limited liability companies, local government units, excluding provinces). In this case, the LAG is overseen by the provincial marshal. If there is provision for this in the statutes, an association may carry out economic activities for the implementation of the LDS. An analysis of the operation of LAGs in Poland leads to the conclusion that the commercial operation of LAGs constrains and raises doubts about the participation of representatives of local authorities in the bodies (especially individuals who hold local government functions and are legally prohibited from conducting business). This is a hitherto unresolved issue in relation to the participation of local authorities in the LAG structure.

In the case of the analysed LAG "Kraina Mlekiem Płynąca", "The Land Flowing with Milk", a foundation was adopted as the legal structure. The regulatory document is the statute, which contains the legal arrangements that form the basis for the operation of rural development. The named LAG has a statute and elected authorities: council, board of directors and an audit committee. Its most important decision-making body is the members' assembly. Public sector representatives may not comprise more than 50% of the LAG board.

1.5 Local Action Group "The Land Flowing with Milk" – the process of building local partnership and bottom-up Leader initiative on the example of the municipality of Mały Płock in the Podlaskie province

The Local Action Group "The Land Flowing with Milk" is located in the municipality of Mały Płock. This municipality is one of the local authorities which, since

Poland's accession to the European Union, i.e., since 2004, has participated in the implementation of the LEADER+ initiative in Poland within the framework:

- "The Pilot Programme Leader+ Scheme I and II, funded under the Sectoral Operational Programme "Restructuring and modernisation of the food sector and rural development 2004–2006",
- Leader Axis 4 of the Rural Development Programme 2007–2013, and measures under the LEADER Rural Development Programme 2014–2020.

The municipality of Mały Płock is a typical rural municipality in north-eastern Poland, located in the Kolno district of Podlaskie Province. The creation of a local partnership in the municipality of Mały Płock was initiated in 2003 with the support of a specialist from the EURO-NGO programme. The programme was financed by the Polish American Freedom Foundation and implemented by the NGO Support Network SPLOT, in cooperation with the Stefan Batory Foundation in Warsaw and the Polish NGO Representation in Brussels. The stimulus for the development of the LAG initiative in the municipality of Maly Plock was the participation in the "Workshop for an Idea" competition organised as part of the "Internet in Schools – Project of the President of the Republic of Poland" programme. For its involvement in the project and for organising information campaigns on Poland's accession to the European Union, the Pope John Paul II Grammar School in Mały Płock was awarded the main prize of PLN 100 000. The outcome of winning the competition was the establishment of the Community Centre for Supporting Local Initiatives of the Municipality of Little Plock. This was established with the backing of rural community leaders, teachers, entrepreneurs, and local government officials. It started the process of community activation to build local partnerships, as an informal organisation. The objectives set were pursued through socio-economic development projects for which both European and private funding was applied for.

Developed and implemented projects were initiated by the Social Centre for Supporting Local Initiatives of the Municipality of Maly Plock in cooperation with the:

- Mały Płock Municipality,
- Gimnazjum im. Papieża Jana Pawła II in Małym Płocku,
- Primary schools in the municipality,
- Volunteer Fire Brigade of Mały Płock,
- Social Welfare Centre of Mały Płock,
- Municipal Police,
- local businesses,
- Co-operative Bank in Kolno,
- agricultural advisory centre District Office in Kolno,
- District Police Headquarters,
- State Fire Brigade,
- District Employment Office,
- local associations,

identifies the stages of development of the partnership, the areas of activity pursued and sources of funding. The projects implemented by the local partnership in the municipality of Mały Płock were the basis for the creation and development of the Local Action Group "Kraina Mlekiem Płynąca". These included:

- "Communal Information Centre" in Mały Płock as part of the "PIERWSZA PRACA" Graduates' Vocational Activation Programme, the Municipal Office in Mały Płock received a grant of 46,300 zloty. This grant was awarded to 13 municipalities in Podlaskie Voivodeship. Total cost of the project: *PLN 66.1 thousand.eVITA programme "Wieś @ktywna – Building an information society" Project implemented in the municipality of Mały Płock.* Contributors: Rural Development Foundation, Polish-American Freedom Foundation and Cisco Systems Poland. Total value of grants – 352, 400 zł.

Moreover, subsidies were extended by the Small Grants Fund:

- "Online Embroidered World" submitted by the Municipal Cultural Centre in Little Plock, 5,000 PLN,
- "A virtual tour of Kurpie culture" submitted by the European School Club "Europejski Krąg" from Mały Płock, 5,000 PLN,
- "Z angielskim na ty – laboratorium językowe w Małym Płocku" submitted by the Foreign Language Club at the John Paul II Grammar School in Mały Płock, 5,000 PLN.
- "An online map of dangerous places – safe road to school", submitted by the group "Safe on the road", operating at Chludnie Primary School, 5,000 PLN.
- "Sports and recreational activities for schoolchildren – With sport for you". Project funded by: Ministry of Education and Sport, Department of General Sport. The municipality of Mały Plock received a grant of 12,000 PLN for the task, representing 50% of the qualifying costs. Total cost of the project: 24,000 PLN.
- *Social Integration Club in Mały Płock – As one form of social employment* Project financed by the Ministry of Social Policy. The municipality of Mały Plock received a grant of 15,000 zł for the task, representing 50% of the qualifying costs. 30,000 PLN.
- "Development of the activities of the Municipal Information Centre" in Mały Płock within the framework of the Programme for Professional Activation of Graduates "PIERWSZA PRACA" in the competition conducted by the local government of the Podlaskie Province Project implemented by the Provincial Labour Office in Białystok. The municipality of Mały Plock was one of only four in the province to receive the maximum funding of 25,000 PLN for the task, representing 70% of eligible costs. Total cost of the project: 36,160 PLN.
- "Safe Municipality" Competition announced by the Ministry of the Internal Affairs and Administration, the Municipality of Mały Plock won first place in the category of rural municipalities and was awarded 50 thousand PLN.
- Computer Labs for Schools in 2006. Primary Schools, Lower Secondary Schools Realised: Chludnie Primary School as part of the programme "Providing support

for the development of the employment sphere by promoting behaviour that contributes to employability, the conditions for equal opportunities entrepreneurship and investment in human resources". Project value 80,000 PLN.

– *eVITA 2 programme "Wieś @ktywna – Building the information society"*. Project implemented by the Land of Milk Flowing Foundation. Contributors: Rural Development Foundation, Polish-American Freedom Foundation and Cisco Systems Poland. Total value of grants – 56 thousand zł. Projects included: Non-commercial Computer Network, Public Internet Access Point, Volunteer Fire Brigades fire safety portal, 'Green virtual map – safe holidays'.

– *"Preschool Clubs in the Countryside"*. Partnership project implemented by FWW (Rural Support Foundation) in cooperation with the J.A. Comenius Foundation for Child Development, the District Teacher Training Centre in Lublin and the Teacher Training and Continuing Education Centre in Suwałki. Comenius, the District Teacher Training Centre in Lublin and the Teacher Training and Continuing Education Centre in Suwałki. The project is funded by the European Union under the framework of the European Social Fund. Value of the project 95 thousand zloty, the project in the Municipality of Maly Plock is handled by the Municipality Office.

– "Municipal Social Integration Programme" within the framework of the Post-Accession Rural Support Programme, a programme implemented by the Municipality of Maly Plock. Project value 70,500 euros.

– "School on the Eye" within the framework of the Government Programme implemented by the Primary and Middle School in Little Plock. Project value 27 thousand zł, including a grant of 18,400 PLN.

– My sports field "Orlik 2012", as part of a government programme, is a project to build a sports complex in the municipality of Mały Płock. Project value 1.03 million zlotys including external funding 666,000 zlotys. This included 333 000 thousand zlotys from the Podlasie Regional Government and 333 zlotys from the Ministry of Sport.

The experience of developing competition projects has significantly improved the qualifications of the individuals and leaders involved in the work of the Social Centre for Supporting Local Initiatives of the Municipality of Mały Płock. When the activities under the SOP "Agriculture" were launched, the local partnership from Mały Płock, which included the commune government, started preparing and applying for preparatory funds to build and prepare for operation of the LAG. This resulted in a project to which neighbouring municipal authorities, i.e., municipalities, were invited: Nowogród, Grabowo, Zbójna, Turośl was the establishment in 2006. Local Action Group "Kraina Mlekiem Płynąca", which became a legal foundation. The project was funded through the Sectoral Operational Programme 'Restructuring and modernisation of the food sector and rural development 2004–2006'. The implementing authority was the Foundation for Agricultural Assistance Programmes – FAPA. The municipality of Mały Płock, as the leader of the Local Action Group, was the only local authority in the province to receive a subsidy amounting to 147.2 thousand zlotys for project implementation, covering 100% of the eligible costs. The implementation area included five partner municipalities: Mały Płock, Grabowo, Nowogród, Turośl, Zbójna. The headquarters of the LAG was a historic manor house from the 19th century, which was also refurbished as part of the local partnership's activities.

The sponsors of the Local Action Group "Kraina Mlekiem Płynąca" were individuals gathered at the Social Centre for Supporting Local Initiatives of the Municipality of Mały Płock, mayors of municipalities and representatives of a number of private, non-governmental and public entities such as the:
- Rural Development Foundation,
- District Employment Office,
- Co-operative Bank in Kolno,
- Association for the Support of Education and the Labour Market in Łomża,
- Kolno Association of the "Family",
- Agro-Group association for sustainable development,
- Farmer's House,
- Apro-Group Development Programming Agency,
- Kolno District Agricultural Advisory Team (PODR Szepietowo).

Supporting entities were the:
- Kolno District Fire Brigade Headquarters,
- Kolno District Police Headquarters.

Two neighbouring local administrations, the municipalities of Kolno and Stawiski, were invited to join the growing LAG. From the outset, the project envisaged institutional support. There were a number of community activation and investment projects involving the development of technical documentation, and investment feasibility studies for preparing investments to benefit the local community (table 3.1).

As a result of its activities, the Local Action Group "Kraina Mlekiem Plownąca" entered competitions under the Local Development Strategy under Axis 4 of the Rural Development Programme 2007–2013. In doing so, it performed the function assigned to the LAG in implementing the LSR by calling for proposals and selecting activities. The LAG has also expanded its catchment area to include the town of Kolno. As a result, the group incorporates an area of 10 municipalities – eight municipalities in the Kolno district and two municipalities in the Łomża district. The Local Action Group "Kraina mlekiem Płynąca" (The Milk-Flowing Land) implements tasks resulting from the agreement signed with the Podlaskie provincial government on the implementation of the Local Development Strategy within the framework of the LEADER Programme for Rural Areas Development 2014–2020. In accordance with the applicable legislation, it adopted the legal form of an association, adapting its structure to the current legislation aligning the possible legal forms of LAGs in Poland.

TABLE 3.1   A summary of implemented public-private projects undertaken in partnership with
the municipality of Mały Płock showing the total amount of outlays incurred and the
share of external funding (private, public, EU) in the period 2003–2010 when the Local
Action Group was established

Project title / programme	Origin and source of funding (private/public/EU)	Total expenditure / PLN	Aid amount / PLN
Social Centre for Supporting Local Initiatives of the Municipality of Mały Płock	Private	120,000	100,000
"Community Information Centre" in Mały Płock	Public	66,100	46,300
eVITA programme "Village @active – Building an information society".	Private	452,400	352,400
"Sports and recreational activities for schoolchildren – With sport for you".	Private	24,000	12,000
Social Integration Club in Mały Płock – As one form of social employment	Public	30,000	15,000
"Development of the activities of the Municipal Information Centre" in Mały Płock	Public	36,160	25,000
"Safe Community".	Public	50,000	50,000
Computer Labs for Schools 2006. Primary Schools and Junior High Schools	Public, EU	80,000	80,000
eVITA 2 programme "Village @active – Building the information society".	Private	56,000	56,000
"Pre-school clubs in the countryside".	Public	95,000	95,000
"Community Social Integration Programme".	Public	285,000 (approximate value)	285,000 (approximate value)
"Szkoła na Oku"	Public	18,400	18,400
My Playing Field "Orlik 2012"	Public	1,030,000	666,666

TABLE 3.1   A summary of implemented public-private projects (*cont.*)

Project title / programme	Origin and source of funding (private/ public/EU)	Total expenditure / PLN	Aid amount / PLN
"The Land Flowing with Milk" in the framework of Measure 2.7 "Pilot Programme Leader+", Scheme II.	Public, EU	147,200	147,200
"Restoration of a historic manor house and farm building, including landscaping" under the activity "Village renewal, conservation and protection of cultural heritage".	Public, EU	526,600	279,000
"The Land of Flowing Milk" in the framework of Measure 2.7 "Pilot Programme Leader+", Scheme II. *Full project title: "Improving the quality of the rural population's life by supporting the activities of the Local Action Group -Foundation "The Land Flowing with Milk" to implement the Integrated Rural Development Strategy developed for the LAG's sphere of action "*	Public, EU	444,400	444,400
Total	Private	3,461,260	2,672,366
	Public, EU	508,400	508,400
	(including local and central government)	2,952,860	2,163,966
Level of external fund commitment	Total expenditure	100%	77.21%
	Private funding	14.69%	19.02%
	Public funding	85.31%	80.98%

SOURCE: AUTHORS' CALCULATIONS BASED ON PUBLIC INFORMATION

Outlining the context of the activity of the Local Action Group "Kraina mlekiem Płynąca", it is necessary to point out conclusions which are important for the implementation of model solutions resulting from the protection of water resources within the territory of the European Union. Firstly, Local Action Groups, as a specific form of NGO, play an important role in shaping and developing rural areas. Secondly, Local Action Groups, acting on the basis of the local partnership principles set out in the LEADER programme, are an important formal element of building local private-public-social partnerships Thirdly, the legal, organisational and financial framework of the Local Action Groups, by being anchored in the structural policy of the European Union, provides a long-term perspective for their institutional development. Fourthly, LAGs actively participate in the process of decentralising decision-making in the socio-economic development of rural areas by integrating local communities around common projects. Fifthly, in assessing the usefulness of local partnerships for the protection of water resources in the European Union area, the example of its activities in the municipality of Mały Płock shows that it is highly effective, not least in obtaining external funding for the implementation of local community support projects (the share of external funding in the volume of total expenditure incurred for the implementation of projects until 2020 was 77.21%). Sixthly, the share of private funds in the breakdown of total LAG expenditure on projects was 14.69%, while the share of private funds in the breakdown of externally raised funds for projects was 19.02%. Seventhly, an in-depth analysis of the effectiveness of the application of the LEADER method in European Union countries other than Poland was also carried out and the achievement of socio-economic effects adequate to those presented in the case study was confirmed. Examples of the LEADER method used by LAGs from the Podlasie region include:

- Belgium – Brussels, Study tour on the programming and implementation of projects implemented under the Structural Funds of the European Commission's Directorates-General for Agriculture and Regional Development and the representatives of the Federal German state of Brandenburg in Brussels. *"Study Tour to Brussels on Agriculture and Rural Development"*. Project implemented by the Ministry of Agriculture and Rural Development and the Deutsche Gesellschaft fur Technische Zusammenarbeit (GTZ) GmbH Delegating authority – Marshal's Office of the Warmińsko-Mazurskie Province, (31.03–06.04.2001).
- Estonia, Latvia, Lithuania, Russian Federation (Kaliningrad Oblast), Participation in a *study tour on establishing cooperation with NGOs and local authorities working for the development of tourism and ecology in the Baltic countries* (Matsalu National Reserve, Pape Lake, Kurskaya Kosa). Project financed by the Royal Netherlands Embassy Delegating institution – Living Architecture Studio Association in Goniadz (16–22.06.2003).
- Finland (Turku), Participation in the "III Baltic Sea NGO Forum". Project implemented by the Russian-Finnish Society. Recruitment Competition – Regional Information and Support Centre for Non-Governmental Organisations

Foundation in Gdańsk! Organised by the Association Pracownia Architektury Żywej in Goniadz: Baltic Sea State Council, (08–11.05.2003).

- Germany (St. Marienthal, Ostritz), International Meeting Centre, Seminar: Opportunities for European Union support for the environment, Project implemented by the Saxon State Foundation for Nature and the Environment (Sachsische Landesstiftung Natur und Umwlet) Delegating Institution – Living Architecture Studio Association in Goniadz, Marshal's Office of the Podlaskie Province, (23–28.11.2003).
- France (Cap Corse – Corsica) Participation on behalf of the ""Kraina Mlekiem Płynąca" Foundation in the Leader+ Observatory Seminar "The Legacy of Leader+ at local level: Building the future of rural areas" Seminar of the European Leader+ Observatory "Leader+ legacy: building the future for rural development" Seminar funded by the European Commission, (23–27.04.2007).
- Portugal (Evora), Participation on behalf of the Foundation ""Kraina Mlekiem Płynąca" in the concluding Conference of the European Leader+ Observatory "Leader achievements: a diversity of territorial experience" Seminar financed by the European Commission, (22–23.11.2007).
- Spain (Aragon and Castilla La Mancha), Participation on behalf of the Podlaskie Agricultural Advisory Centre in Szepietów in a study and training trip on "LEADER in Spain – review and dissemination of good practice" tourism-information technology – rural business development. Trip organised by the Polish-Spanish Cooperation and Development Foundation in cooperation with the Department of Promotion and Trade at the Polish Embassy in Madrid, (01–06.12.2008).
- Ireland (Tipperary), Participation on behalf of the "Kraina Mlekiem Płynąca" Foundation in a study trip to the Tipperary Institute, trip financed under the "Leader+ Pilot Programme" Scheme II, (03–07.03.2008).

Many years of involvement of the entities of the Local Action Group "Kraina Mlekiem płynąca" ("The Land of Flowing Milk") in the implementation of the LEADER method makes it possible to confirm that the implementation of solutions for the protection of water resources in the area of the European Union is possible. This proposal can be further extended to include LAGs operating in all countries of the Community. This study is based on research contained in the publication by S. Gromadzki and N. Chodkowska, on the local action group as a specific form of NGO operating in rural areas.[7]

---

7 Sławomir Gromadzki, Natalia Chodkowska, Lokalna grupa działania jako specyficzna forma organizacji pozarządowej działającej na obszarach wiejskich – proces tworzenia partnerstwa na przykładzie gminy Mały Płock, in Organizacje pozarządowe w ujęciu prawno-postulatywnym, Urszula Szymańska, ed. Monika Falej, Piotr Majer (Olsztyn: Wydział Prawa i Administracji UWM, 2017), 205–221.

2   **Applicability of the LEADER method in the process of establishing a local water partnership – theoretical assumptions**

The potential for adopting the LEADER method for the protection of water resources in the European Union's territory stems primarily from the participation of the local community in the creation and implementation of a local strategy for the protection of water resources, particularly water retention. This relates to a number of key factors determining the feasibility of the concept: bottom-up, territoriality, integrity, partnership, innovation, decentralised management and funding, network centrality and partnership and in addition, the participation of diverse economic actors – businesses, local authorities, and community organisations.[8]

2.1   *Bottom-up*

The effectiveness of the LEADER method in the implementation of local projects makes it possible to structure a "local water partnership" model based on existing solutions and experience gained over time in the European Union and its Member States. In the proposed solutions, this kind of partnership is formed as a bottom-up initiative of social, public, and economic actors. The public sector, as appropriate for the LAG, is represented by local authorities, district agricultural advisory centre teams, other public entities and their branches operating at local level, fire brigade units, and emergency response teams. In the proposed solutions, the economic sector is represented by water companies, farmers, businesses interested in the area of activity of the 'local water partnership'. The assumption is that they and furthermore community associations will be the most willing creators of proposed solutions. The bottom-up initiative of the above-mentioned stakeholders is organised on the basis of the applicable principles and rules expressed in legislation at both national and EU level. This should function as it does for local action groups. It is intended that the indicated means of organisation will support the development of local entrepreneurship and innovation.[9]

2.2   *Territoriality*

The area of action of the "local water partnership" appropriately for the analysed Local Action Groups defines a coherent administrative area (this applies to all or part of the local authorities that have joined the partnership). In the proposed solutions, unlike in the case of LAGs, it is assumed that in the case of the 'local water partnership' model, a municipality can be a member of several 'partnerships'. This

---

8   Kim Pollerman, Processes of cooperation in rural areas: obstacles, driving forces, and options for encouragement, rural cooperation in Europe, ed. Edward Kasabov (Basingstoke: Palgrave Macmillan. 2014), 210–227.

9   Lois Labrianidis, "Fostering entrepreneurship as a means to overcome barriers to development of rural peripheral areas in Europe," European planning studies 14 no. 1 (January 2006): 3–8, https://doi.org/10.1080/09654310500339067.

is because the basis for water protection and retention planning is primarily the hydrological division of an area (catchment areas). The administrative area of a local authority may include several catchments or parts of catchments. In practice, this means that the local authority area may overlap with several geographically distinct water catchments. The actual needs and desirability of establishing a 'local water partnership' will be determined by conditions: soil, geomorphology, hydrology and hydrogeology, relating to the profitability of agricultural production, the protection and restoration of the natural environment and the probable occurrence of water deficits, relating to predicted climate change. From a functional point of view, it is assumed that the local government can function in several 'water partnerships' representing individual villages, districts, towns and areas.

2.3   *Internality*

Most important in achieving the functional objectives of 'local water partnerships' is the integration of different socio-economic fields, i.e., the coherent combination of the interests of local business, agriculture, local government bodies and NGOs. Integration makes it possible to achieve a synergistic effect, i.e., complementing the activities of individual entities, sectors of the economy, thus increasing its effectiveness and increasing the effects and profitability of the projects undertaken. Potential benefits include:
- Determining the optimum location and required capacity of retention facilities and reservoirs.
- Demonstrating the need for the construction of new reservoirs or their modifications, reconstruction/rebuilding of existing culverts, weirs and other hydraulic and drainage infrastructure.
- Support for the provision of connection conditions to the existing primary and secondary drainage network.
- Mapping of overloaded or under loaded sections (especially in the context of simultaneous use of the system for flood protection).
- Optimal extension planning of retention and flood control networks based on local spatial plans or other planning guidelines of municipalities, in this case municipal strategies and plans for water protection and water retention.
- Agricultural water supply strategy for agricultural production (irrigation).
- Planning of repairs based on failure history analysis of a section of the network.
- Reducing threat response times.
- Network centric process management.
- Real-time control (RTC) of water resource protection, and retention, at the system management level.

In addition, the proposed solutions preserve their autonomy (achieving the partnership members' particular objectives) and overarching objectives (impossible for individual partners to achieve). Integrating activities also allows a compromise to be found in the case of equal / opposing business objectives / interests (e.g., that of operators, farmers, and environmental organisations). This allows us to understand different standpoints and concerns and to develop a coherent strategy of action that includes compromise. In the case of water management, the trade-off between resource exploitation and conservation is crucial, as it affects a range of areas, from the European Union's climate policy to the economics of green capitalism determined by the fourth industrial revolution.

2.4   *Partnership*

Central to the proposed solutions is local partnership, involving various stakeholders from the public, social and economic sectors.[10] The solutions model assumes that the 'local water partnership' takes a specific legal form as a special association. Partnerships can include:

- natural persons (farmers, sole traders),
- legal persons including associations, foundations, limited liability companies, cooperatives, water companies, and producer groups,
- unincorporated organisational entities, e.g., district fire brigade headquarters, police headquarters, municipalities, social welfare centres, emergency response teams and others.

Entities may participate in partnerships represented by their representatives on the basis of a power of attorney. The initiative to establish a 'local water partnership' is held by all of the above-mentioned stakeholders. The founding group invites other stakeholders and convenes a general meeting, which is announced to the public with the right of other interested parties to join the initiative. In the framework of the general meeting for the establishment of the future partnership, each entity declaring its intention to join has one vote. The general meeting elects a meeting chairman and a returning committee. A draft statute is drawn up and adopted by vote. The voting procedure shall be determined by the general meeting. 'Partnership' constituent bodies such as the partnership council, board of directors, audit committee and supporting bodies e.g., programme board, scientific council, social council and other committees are elected. The adopted model assumes that the local government(s) in the area where the "partnership" operates are obliged to provide assistance in the organisation of the "partnership", i.e., administrative, legal services, indication of the location or premises for the office, etc. This is born of the nature of a 'local water partnership' working for the benefit of the local community, a given geographical region defined by water catchments. After the adoption of the statutes and the election of the 'partnership' bodies and their constitution, the partnership's board files an application to the competent court for registration (in the case of Poland, for example, the National Court Register). Once the 'partnership' is registered, it starts its activities, organises its office, and undertakes the

---

10   Law dated 19 December 2008 on public-private partnership, Journal of laws. 2009 No. 19
       item 100.

implementation of the agreed objectives. The primary objective of the partnership is to undertake the development of a 'local water retention and conservation strategy' as a strategic document defining the scope and timing of the actions to be taken, i.e.:

- analysis of site topography,
- analysis of existing and planned stormwater sewers and reservoir retention,
- verification of hydrological assumptions,
- development of a concept and functional programme for water protection and retention,
- development of assumptions for a hydrodynamic model for water protection and retention,
- the development of guidelines for the construction of model rainfall hyetographs and any elements enabling the use of intelligently controlled reservoir retention systems for water conservation.

Adoption of intelligently controlled reservoir retention systems for water protection will enable:

- reduction of economic and environmental losses.
- monitoring of important operational parameters necessary to control the flow and pressure of water in the network.
- ongoing data processing, archiving and internal communication within the system.
- real-time visualisation of basic retention processes, including maximising the time between the occurrence of an event (e.g., failure) and the notification of its occurrence to the operator
- remote access to the system to identify potential threats and respond to them immediately.

Having the inputs identified above to formulate the strategy will allow it to be adopted by the Partnership Board. Then the entity called "Local Water Partnership" presents the developed and adopted "Local Water Retention and Protection Strategy" to the local government, and the cooperating local/county governments in the area in which it operates. The document is accepted or rejected for implementation. In the latter case, it is returned for reworking with comments. In the other case, the district/municipal councils accept it, and it becomes local law defining the nature of the activities of the 'Local Water Partnership'.

2.5   Innovation

The innovativeness of the solutions associated with the establishment and operation of the 'Local Water Partnership' entity stems from the implementation of the 'Local Water Retention and Protection Strategy'. It is intended to provide a platform for the development of social and technological innovation. In the designed solutions, the continuous improvement of competences and qualifications will enable added value and create awareness. In the former area, the focus is on value creation and value capture and above all identifying individual and collective needs, preparing products, offers, modes of action, identifying and classifying data. Against this backdrop, the capture of value takes the form of the sale of products/items, the production of goods, services, the use of existing resources and procedures in terms of development capacity.

In the second area, it is about the immediate response to individual and collective needs, the continuous improvement of offers, products, the convergence and use of information ensuring the improvement of existing products/offers and the creation of new ones, the duplication of income, a personalised network-centric approach and the creation of new competences. The innovativeness of the action of the "Local Water Partnership" entity is expressed through the management and design methods used and the techniques and technologies implemented in the area of water retention and protection.

The economic benefits for local communities seem indisputable – from strengthening the resilience of the natural environment (biodiversity), to its attractiveness for tourism (strengthening tourism in specific regions), to the economy of the entire European Union and the entire organisation – from agriculture, to new technologies (communications, sensors), the implementation of economy 5.0 at local level, above all in the form of human interaction and artificial intelligence, environmental protection through the use of renewable energy (sun, wind, water), and the elimination of waste. Innovation can also relate to: technologies used in the agricultural sector, local businesses using water resources – renewable energy, – agricultural drainage, irrigation systems, reclamation, treatment, emergency management, and education, and training. Leveraging the 'Local Water Partnership', it is easy to identify an interpretation in which fields – from sowing to harvesting – and water are continuously monitored, providing an understanding of the nature of crop growth, irrigation needs or pests. It is clear that this element correlates with water resource protection and water retention, which will generate natural, focal point sources of energy supply. Given the nature of the scale – the use of drones and sensors for this – means that they will become as assistive as, say, a combine harvester. The deployment of the internet of things/things for these purposes can be used to monitor the environment (floodplains, sewage treatment plants), collect meteorological and climatic information, track animals including their permanent and periodic migrations.

2.6   Decentralisation of management and funding

In the proposed solutions for the protection of water resources and their retention, in the territory of the European Union, it is assumed that the "Local Water Partnership" has guaranteed funding from the state budget and targeted operational programmes implemented in the member country and managed at regional level. Analogously, as has been described in the example of the activities of LAGs in Poland under the Leader initiative in the Rural Development Programme managed at regional level by the Provincial Government. It is assumed that the 'Local Water Partnership' operates on a defined budget based on a population indicator (as appropriate for the LAC) and an algorithm adopted by the European Commission taking into account hydrological conditions, surface area, risk of natural disasters (droughts, floods) and the level of environmental contamination.

Within the agreed budget, the "partnership" benefits from a set rate of 10–15% for administrative, training and other indirect costs. The budget for the targeted actions of the "partnership" is set in accordance with the programming periods of the European Union Structural Funds, hypothetically for the years 2021–2027 for the tasks enshrined in the "Local Water Retention and Protection Strategy". Funds are to be disbursed through competitions. Accordingly, as in the LAG, the "Local Water Partnership" would act as the local implementing body responsible for setting the criteria for competitions, launching competitions, selecting operations, establishing ranking lists, and forwarding the ranking lists to the competent authorities at regional level for the signing of contracts. In addition, the 'Local Water Partnership' would participate in the process of clearing and controlling the implemented activities. Synergies are key to achieving the objectives stipulated in the 'Local Water Retention and Protection Strategy'. In principle, this takes place by pooling public and private funding. This maximises the effectiveness of fund expenditure and achieves economies of scale across countries and the European Union as a whole.

2.7   *Network centricity and cooperation*

The adoption within the European Union of the proposed 'Local Water Partnership' creates economies of scale enabling the exchange of experience and dissemination of good practice between its member states. Here, it is important to identify three levels on which this interaction can take place. Firstly, at the level of local partnerships within shared catchments where there are similar climatic and hydrological conditions, common problems and risks of natural disasters, and similar environmental and economic conditions. Correspondingly, like the LAGs operating within the organisation, the proposed cooperation model assumes network centric partnerships at local level. This is particularly relevant for water catchment areas that typically flow through different countries of the European Union. Existing solutions within the Ecrins framework make it possible to exploit full topological information within a geographical information system of European hydrographic systems in the activities of the 'Local Water Partnerships'. This is particularly important for the functioning of local partnerships within defined catchment areas. Secondly, at the regional level, using Ecrinis and the synergy of the various regions of the European Union, it is possible to establish joint pro-environmental measures. Regional cooperation activities can include:
– established organisations bringing together 'Local Water Partnerships',
– implementation of joint training, exchange of experience and know-how in the field of water protection and retention,

- consultation on problems and lobbying activities,
- developing joint legislative proposals.

This creates an opportunity for the European Committee of the Regions (CoR) to introduce a new cooperative formula focusing on water protection and water retention, involving individual citizens, local stakeholders, local communities and regional authorities. Thirdly, at European Union level, the Eco-Management and Audit Scheme (EMAS) should be an additional tool to support the activities of 'Local Water Partnerships'. Its application in assessing, reporting and improving the environmental performance of local partnerships in reducing energy consumption, waste generation and greenhouse gas emissions would be particularly valuable in assessing the environmental impact, climate policy of organisations. Leveraging Ecrinis and EMAS in the activities of 'Local Water Partnerships' will enable:

- linking a range of innovation initiatives to the climate policy objectives of the European Union,
- targeting local, regional and institutional cooperation in the European Union in areas not yet covered by its environmental policies, which are linked by the formula of eco-capitalism to the economy, the quality and lifestyle of local communities,
- the multifaceted cooperation of European Union countries and institutions with each other.

In addition, the inclusion of 'Local Water Partnerships' in institutional cooperation will allow the designated European Union institutions to take an even more comprehensive approach to environmental protection, the fight against climate change, the green economy, the involvement of communities and local stakeholders in its policies, and the ways in which they are implemented. There is a need to introduce water retention and drought prevention throughout the European Union. and moreover, involving community representatives and local residents through self-identification.[11]

3   Conclusions

The nature of the changes that can potentially emerge in the European Union through the inclusion in its policies of a 'Local Water Partnership' as the entity that programmatically implements them in every community of every Member State is worthy of special attention. By introducing the described formula for water protection, including water retention, it becomes possible to create an interpretation

---

11   Stefano De Rubertis, Foreword, in Neo-endogenous development in European rural areas results and lessons, ed Eugenio Cejudo, Francisco Navarro (Springer International Publishing 2020), V–XI.

combining protection of the environment, development of the green economy, improvement of living standards and adoption of pro-ecological forms of behaviour by local communities, entrepreneurs, and farmers. The adoption of LAGs as a basis for creating "Local Water Partnerships" adapts existing European Union operating instruments. Moreover, it brings together Ecrinis, EMAS, and the European Committee of the Regions in a synergistic way. This creates the conditions for the success of the described project establishing "Local Water Partnerships" at Member State level and their use in EU policies aimed at protecting environmental resources, biodiversity, the green economy, and combating global warming.

In the case of the use of the 'Local Water Partnership' as a European Union policy measure, economies of scale are to be found. This relates to the implementation of activities, and their effectiveness in achieving objectives. The matter concerns institutional, systemic, complex, infrastructural, environmental, economic, social and political solutions. In each of these, the scale of the interconnected activities is pan-European, although the extent to which the intended solutions are applied is likely to be substantially different in each region of the EU. The concept of water protection and water retention, points to the need for a holistic approach, with synergistic action by many institutions and actors, including farmers/businesses and the public. An information and education campaign are a prerequisite for ensuring success. Above all, it should be delivered at a local level, targeting the local community and farmers/entrepreneurs. The involvement of the latter, together with local authority institutions, is key to the success of the whole concept. The consequences will be the expansion of the infrastructure of the fourth and fifth industrial revolutions, as well as environmental, economic and security benefits for societies and countries in the European Union.

The use of the "Local Water Partnership" as an entity to ensure water resource protection, as well as water retention, makes it possible to reduce the costs of environmental and economic measures through, among other things, more effective planning and investment efficiency. The economic benefits appear indisputable. These include:

- strengthening the resilience of the natural environment (biodiversity),
- strengthening agriculture, local enterprise, and local communities,
- the attractiveness of the regions for tourism (strengthening this sector),
- the development of new technologies dedicated to Industry 5.0 at local level, primarily in the form of human interaction and artificial intelligence, and the Internet of Things by combining digital and physical devices (e.g., water, soil, air sensors),
- the elimination of waste and the use of environmentally friendly technologies that reduce the carbon footprint,
- economies of scale across the European Union.

In the adopted assumptions, the European Union through "Local Water Partnerships" will gain new capacities in the form of:

- immediate provision of anticipated environmental, economic, and social needs,
- convergence of information from different sources in the existing Ecrinis and EMAS systems,
- dedicated information to meet the needs of farmers, entrepreneurs, local communities and provide appropriately configured information to Member State institutions and organisations,
- improving policies, for example the Common Agricultural Policy, the European Green Deal, with the involvement of the Union's institutions, in particular the Committee of the Regions.

Making use of strengths such as a protected and enhanced environment, reduced cost-intensity of agricultural products and others, derived from this, allows the European Union to exploit its development potential, creating a competitive advantage for its Member States. Cooperation of the European Union with other countries, political players, in the identified area of activities would allow strengthening and creating additional political capital, internally and in international relations, for example in relations with the Food and Agriculture Organisation of the United Nations.

The protection of water resources, as well as water retention using pan-European 'Local Water Partnerships' will enable the European Union to gain comparative advantages. Additionally, it will allow the shaping of new capacities already named, as well as further capacities related to the organisational and strategic culture. The involvement of the European Union will create space for the creation of interdependencies between individuals, social groups, society and public institutions. The use of the presented model of interaction between different private and public stakeholders will allow a more effective implementation of the European Union's policies for social mobilisation related to the achievement of climate, economic, and social objectives. In practical terms, the implementation of "Local Water Partnerships" within the implementation of European Union policies makes it possible to configure a consolidated strategy for social and economic development with a focus on protecting and enhancing environmental biodiversity, and economic efficiency of agriculture. A crucial determinant is the coherence of activities together with their bottom-up nature. Given the climate policy goals of the European Union, namely the shift away from the use of carbon and its reduction in the atmosphere, the timing of the 'Local Water Partnerships' activity is crucial. Given the climate policy goals of the European Union, namely the shift away from the use of carbon and its reduction in the atmosphere, the timing of the 'Local Water Partnership' activity is crucial.

CHAPTER 4

Local water partnership

The aim of this chapter is to formulate assumptions and methodologies for the creation of 'local water partnerships' as the basic organisational unit (public-private-nongovernmental partnerships) performing the task of planning and implementing 'local water retention and protection strategies at the local, micro-catchment (micro-catchment) level. Under the cognitive assumptions made, only this type of measure directed at the protection of water resources was identified as desirable, producing a specific result for the environment, humankind and agricultural activity.

The United Nations General Assembly, in its resolution A/RES/47/193 dated 22 December 1992, declaring 22 March each year as World Water Day, draws attention to the global problem of the undervaluation of water resources. Social and economic activities depend heavily on the quantity and quality of fresh water supplies. Water protection and sustainable management requires action at local, regional, supra-regional, individual EU country and community level.

The irrational, wasteful and often damaging use of environmental resources, especially water resources according to the concept of anthropological influence on irreversible climate change,[1] is something that must be arrested for the good of humanity as well as the planet as a whole. To achieve this, the decline in natural freshwater retention capacity must be curtailed and then reversed. Its successive loss affects the state of the environment and living conditions for humans and animals. This generates global implications for the international and national security of individual states, their food, ecological, environmental, social, economic and consequently political dimensions.

For many countries, especially those on the European continent, there has been little consideration for water as it has always been plentiful supply. The situation was similar in Poland, as reported, among others, by the former Minister of Agriculture, Jan Krzysztof Ardanowski. *"Up to now, we have rather been faced with the problem of the permeability of drainage systems, in the event of excess water. Nobody expected that we would live to see a time when water would be in short supply."*[2] In Polish legislation,

---

1   Colin N. Waters, Jan Zalasiewicz, Colin Summerhayes, Anthony D. Barnosky, Clement Poirier, Agnieszka Gałuszka, Alejandro Cearreta, Matt Edgeworth, Erle C. Ellis, Michael Ellis, Catherine Jeandel, Reinhold Leinfelder, J.R McNeill, Daniel deB Richter, Will Steffen, James Syvitski, Davor Vidas, Michael Wagreich, Mark Williams, An Zhisheng, Jacques Grinevald, Eric Odada, Naomi Oreskes, Alexander P. Wolfe, *The Anthropocene is functionally and stratigraphically distinct from the Holocene*, Science (2016), 351. Doi: 10.1126/science.aad2622.
2   Renata Struzik, Tworzone są Lokalne Partnerstwa ds. Wody PGW Wody Polskie, accessed 10.10.2022, teraz-srodowisko.pl.

© JAROSŁAW GRYZ AND SŁAWOMIR GROMADZKI, 2024 | DOI:10.3920/9789004699069_006

the aim of managing water resources is defined as follows: *"(...) The management of water resources serves to meet the needs of the population and the economy and to protect the waters and the environment associated with those resources (...)"*.[3]

1   Current planning status – drought mitigation, and water scarcity using Poland, a central European Union country, as an example

The projected development of the European Union's New Green Deal in the area of agriculture requires the formulation of proposals for solutions which, building on the financial and organisational successes of the programmes and projects implemented in rural areas to date, will achieve the strategic goal of achieving climate neutrality.[4] Only a coordinated effort to protect water resources, to combat drought at the level of the European Commission, the European Parliament and the European Council will make this possible.

The area of rainwater and snowmelt retention in Poland is significantly influenced by climate change which among other things influences the water levels, and their protection.[5] Predicted changes in daily precipitation totals for Poland on the basis of CORDEX simulations indicate that it will rain less frequently and there will be long periods of drought, interrupted by sudden and intense rainfall. Increases in the average daily precipitation totals are forecast to be no more than 10 per cent, meanwhile, the number of days with precipitation exceeding 10mm and 20mm (parameters R10mm and R20mm) in the summer hydrological half-year will increase by a few per cent for R10mm and by more than ten per cent for R20mm. These are area readings for Poland for the year 2050 horizon only, assuming a relatively benign RCP 4.5 climate change scenario. Exceeding the 10 mm depth of precipitation is an essential criterion for it to be considered heavy rainfall, with the potential to result in flooding or the discharge of untreated sewage from the combined sewerage network polluting waters (usually rivers). There will also be a risk that the lack of proper retention and regulation of this rainfall will make cities even drier and more prone to flash flooding and urban flooding during rainfall events.[6]

---

3   Ustawa z dnia 20 lipca 2017 r. prawo wodne, Dz.U.2020.0.310.

4   Sławomir Gromadzki, *Model europejskiego "Lokalnego partnerstwa na rzecz wody". Realizacja celów Europejskiego Zielonego Ładu*, in Zrównoważony rozwój i Europejski Zielony Ład wektorami na drodze doskonalenia warsztatu naukowca, ed. Michał Staniszewski Henryk A. Kretka, (Gliwice: Wydawnictwo Politechniki Śląskiej, 2021), 195–203.

5   Rozporządzenie Ministra Gospodarki Morskiej i Żeglugi Śródlądowej z dnia 11 października 2019 r. w sprawie klasyfikacji stanu ekologicznego, potencjału ekologicznego i stanu chemicznego oraz sposobu klasyfikacji stanu jednolitych części wód powierzchniowych, a także środowiskowych norm jakości dla substancji priorytetowych (Dz.U. 2019 poz. 2149).

6   Rządowy Program Strategiczny Hydrostrateg "Innowacje dla gospodarki wodnej i żeglugi śródlądowej" Projekt założeń, listopad 2021, 11.

Preliminary research conducted for the period 2004–2014 using the example of the Podlaskie Province (North-Eastern Poland) indicates that the share of projects co-financed from EU funds relating to water resources management was only 0.85% in the overall pool of completed projects supported by Measure 2.3 – Village renewal as well as conservation and protection of cultural heritage under the Sectoral Operational Programme Restructuring and modernisation of the food sector and rural development 2004–2006 and by the Measure Renovation and development of villages under the Rural Development Programme 2007–2013.

Based on an analysis of the RDP financial perspective implemented between 2014 and 2020, the number of water management projects accounted for less than 2% nationally. This clearly indicates that the rural development programmes implemented so far, despite their measurable impact on the revitalisation of many areas of the economy, both infrastructurally and socially in the countryside, have not contributed to a structural change in the water protection and retention system.

The applied planning assumptions presented in the chapter on counteracting the effects of drought and water scarcity apply to Poland, an example of a central European country in the European Union. They are intended to complement the current planning model for water resource protection, addressing the effects of drought. Here, a shift in planning methodology to a bottom-up approach in administrative terms, i.e., village – municipality – district – province – state, and in hydrological terms, i.e., micro-catchment – catchment – river basin, has taken place.[7]

1.1 Design for solutions and activities

Currently, the basic planning document for counteracting the effects of drought and water scarcity in Poland is published in the Journal of Laws on 3 September 2021. Plan for counteracting the effects of drought (PPSS) developed by the State Water Management Company Wody Polskie.[8] The implementation of the plan was preceded by:

- the preparation of a forecast of the environmental impact of the draft Plan, together with public consultation on the draft forecast and the draft Plan.
- implementation of the legislative path for the document,
- securing an entry on the legislative work list of the Minister for Climate and Environment.
- completion of intra-ministerial arrangements,
- undertaking external arrangements and public consultations relating to the document,

---

7 Sławomir Gromadzki, Katarzyna Glińska-Lewczuk, Marta Śliwa, "Protection of natural and cultural heritage in rural areas of the Voivodship of Podlasie in 2004–2014 supported with EU funds within the framework of the village renewal contest," Contemporary problems of management and environmental protection, no. 11 (2015): 69–80.

8 The PPSS was drawn up pursuant to Articles 183–185 of the Act dated 20 July 2017. – Water Law.

– receiving opinions from the Joint Commission of Government and Local Government and the Government Legislative Centre.

An analysis of these activities shows that the public consultation lasted from 15 August 2019 through to 15 February 2020. During this period, a mere 15 consultation meetings were organised in 15 cities across the country. Consultation was conducted in a restricted way directly among stakeholders, i.e., farmers, forest holders, local social organisations or local authorities. During the public consultation of the draft, stakeholders were given the opportunity to submit comments and proposals on the draft plan (a total of 850 comments and proposals were submitted during the public consultation of the proposed draft plan, of which 7 anonymous comments were rejected). It should be noted that the development of the plan itself was top-down, and imposed, without the participation and involvement of local partners directly operating in micro-catchment setting, who would actively participate in its development.

The 'Drought Management Plan' document, together with river basin management plans, flood risk management plans and water maintenance plans, is intended to help improve the state of water management in Poland in terms of adequate water quantity and quality for society, the environment, and the economy. *This plan includes:*

1)   An analysis of the potential for increasing available water resources.
2)   Proposals for the construction or rebuilding of water facilities.
3)   Proposals for necessary changes in the way water resources are used as well as changes in natural and artificial water retention.
4)   Measures to counter the effects of drought.[9]

The document includes a catalogue of drought mitigation projects. These are proactive measures to pre-empt the occurrence and reduce the likelihood of adverse effects of drought, implemented regardless of the actual occurrence of a drought event The adaptation approach is intended to promote measures to enhance the characteristics and processes that shape water resources in catchments, for the purpose of reducing losses in the event of a possible drought.[10] These provisions are central to the proposed solutions applying the LEADER method. This method allows for the combination and complementary delivery, at the catchment and river basin area scale, of both technical and non-technical measures to shape water resources, supported by instruments of spatial planning, land and water management, protection of aquatic and water-dependent wetland ecosystems and additionally, the use of instruments to achieve environmental objectives in Poland and within the European Union.

The use of the proposed LEADER-based solutions for drought and flood risk management will contribute to reducing the scale of the risk. The result will be a

---

9   Rozporządzenie Ministra Infrastruktury z dnia 15 lipca 2021 r. w sprawie przyjęcia Planu przeciwdziałania skutkom suszy, Warszawa, dnia 3 września 2021 r. Poz. 1615, 8.
10   Ibid, pp. 8–9.

reduction in water shortages, the strengthening of flood and fire protection, and the improvement of the condition of surface and groundwater bodies. The common denominator for addressing the effects of drought, and floods is the amount of available water resources for use and safeguarding the functioning of ecosystems. Consequently, on a nationwide scale, the implementation of the identified changes in the approach to water retention will contribute to the protection of available surface waters (map 4.1).[11] and consequently to climate change mitigation. A synergy of actions taken to reduce the combined risk of floods, droughts, fires, with appropriate costing, is essential. Among these measures to strengthen and restore the retention capacity of the area should be targeted, including:

1. Protection and restoration of ecosystems.
2. Protection and restoration of biodiversity.
3. *Implementation of the concept of sustainable planning and design of retention in rural and urban areas (so-called smart city, introduction of elements of blue-green infrastructure).*
4. Building a climate-neutral economy.[12]

The solutions adopted in Poland include specific objectives, specifying the main objective of the Drought Mitigation Plan. Above all, in the areas of measures that increase the resilience of vulnerable economic, social and environmental sectors to drought losses, as well as mitigating its effects. These are:

1) Effective water resources management to increase the availability of water resources in river basin districts.
2) Increasing water retention in river basin areas.
3) Education and drought risk management.
4) Funding for drought mitigation measures.[13]

---

11 *Disposable surface water resources 'were determined on the basis of 451 (out of 1212 water gauge stations belonging to PSHM), characterised by complete sequences of daily flow data from the period 1987–2017. For the 117 unmonitored catchment areas, expert methods were used to determine the disposable resources, including interpolation and extrapolation of results. A total of 568 catchments were analysed across Poland. The analysis required the determination of the inviolable flow for each gauge station (QN) in [m3/s] and in terms of outflow modulus [l/s-km2]. The calculation assumes that the amount of water left in river channels as an inviolable flow is determined by the rationale derived from the need to protect the environment and meet the demands of water users. The inviolable flow is thus defined as that quantity of water expressed in m3/s which should be maintained as a minimum in a given river cross-section for biological and social reasons. As part of the analysis, the inviolable flow was calculated on the basis of the hydrobiological criterion (parametric method), known as the Kostrzewa method. The hydrobiological rationale, which determines the preservation of the basic forms of flora and fauna, characteristic of the river's aquatic environment [...], was taken as the basic criterion. The unaffected flow volume values calculated in this way were divided into three classes: those below 2.480 [l/s-km2], those between 2.480 and 4.959 [l/s-km2] and those above 4.959 [l/s-km2]. Intact flow volume values were determined for only 451 monitored catchments'. Ibid, p. 11.*
12 Ibid, pp. 8–9.
13 Ibid.

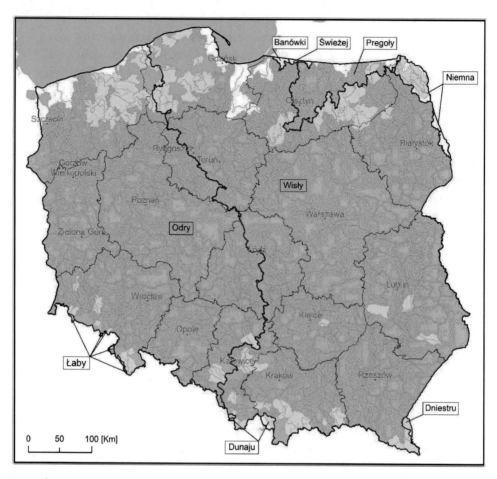

Legend
Intact flow module
[l/s km2]

■ > 4.96

□ 2.48-4.95

■ < 2.48

〜 Polish border

〜 Voivodeship border

〜 River basin areas in Poland (JCWP v8)

〜 Selected rivers (MPHP 10 v8)

■ Lakes and water tanks (MPHP 10 v8)

· Provincial cities

MAP 4.1   Distribution map of inviolable water flow in Poland
             SOURCE: DECREE OF THE MINISTER OF INFRASTRUCTURE DATED 15 JULY 2021 ON THE
             ADOPTION OF THE PLAN FOR COUNTERACTING THE EFFECTS OF DROUGHT, WARSAW,
             3 SEPTEMBER 2021. ITEM 1615, 14

Annex 1 of the *Drought Mitigation Plan* is a list of planned investments and activities. These include investments by the State Water Management Company 'Wody Polskie' (PGW WP) They are intended to contribute to counteracting the effects of drought. These are investments for the development of watercourse retention (small-scale retention), investments by external parties submitted during public consultations, which have passed the initial assessment on the basis of a cost-benefit analysis.

Annex 2 of the Drought Mitigation Plan presents a catalogue of actions whose implementation is expected to contribute to minimising the effects of drought. These include those related to increasing artificial and natural retention, formal and legal actions and educational activities. The main activities related to increasing retention include:

- measures to reduce surface run-off and retain water in nature, in the soil – where precipitation has fallen,
- converting drainage facilities from a drainage function to an irrigation and drainage function, and preserving wetlands,
- the construction of water reservoirs – whether located on watercourses or alongside them.

The main formal measures included those dedicated to the onset of drought and aimed at mitigating its effects, including temporary restrictions on water use. The plan highlights the effectiveness of drought mitigation in the form of complementary measures including:

- technical measures, involving investments in large as well as small-scale retention,
- natural retention, such as re-establishing wetlands,
- increasing riverbed retention,
- non-technical, consisting of developing good mindsets and educating the community,
- building drought monitoring and response systems.[14]

The actions contained in the Drought Plan are complemented by those contained in the Assumptions for the Water Scarcity Programme 2021–2027 with a perspective to 2030. The programme is a response to climate change and rising global temperatures, which are leading to an increase in the frequency of extreme weather events that raise the risk of flooding and drought. *"Currently, Poland stores about 4 billion cubic metres of water in reservoirs, which is only about 6.5 per cent of the volume of average annual river outflow.* Nevertheless, the physical and geographical characteristics of Poland provide opportunities for retention of approximately 15 per cent.[15]

---

14   Ministerstwo Infrastruktury, accessed 4.02.2022, Plan przeciwdziałania skutkom suszy – Ministerstwo Infrastruktury, Portal Gov.pl (www.gov.pl).

15   Water scarcity prevention programme, accessed 2.02.2022, https://www.gov.pl/web/infrastruktura/program-przeciwdzialania-niedoborowi-wody.

"The programme assumes a combination of all available water retention methods: large-scale retention, small-scale retention, artificial retention, natural retention and land reclamation. The appendix containing the adopted programme assumptions includes a list of 94 investments with an assumed value of approximately PLN 10 billion, which are due to be implemented by 2027:
– an increase in the volume retained water.
– an increase in the capacity of small retention facilities.
– drought mitigation with a particular focus on rural and woodland areas.
– reducing the risk of flooding, including that associated with so-called flash floods in urbanised areas.
– restoration or improvement of conditions for the use of water for electricity generation.
– an increase in the contribution of local and regional water retention projects.
– an increase in public awareness of the problem of dwindling water resources and the need for water retention.
– an improved environment for the agricultural exploitation of water.
– enhancement of ecosystems created or maintained through the use of water retention.
– an improvement in the class and stability of navigational conditions of inland waterways.
– an improvement of the landscape qualities of water-related areas.[16]
Implementing Local Water Partnerships in the indicated portfolio of activities creates an opportunity to increase their potential especially in micro catchment areas.

As part of the 2021–2027 programme to address water shortage with a perspective to 2030, investments are planned. Funding for these is expected to come from:
– European funds from the 2021–2027 financial perspective.
– loans or credits granted by international financial institutions: World Bank, Council of Europe Development Bank, European Investment Bank,
– national budget,
– local authority budgets,
– budgets of other entities (e.g., Polish State Waterways, State Forests, National Fund for Environmental Protection and Water Management).
– public-private partnerships".[17]
The configuration of funding from the above-mentioned sources in the activities of Local Water Partnership creates an opportunity to meet the criteria of bottom-up, territoriality, integrity, partnership, innovation, decentralised management and funding, network centricity and cooperation.

An indispensable support for the projected activities undertaken by the Local Water Partnership as part of the Drought Mitigation Plan and the Water Shortage Programme 2021–2027 with a perspective to 2030 is the Drought Monitoring System

---

16   Ibid.
17   Ibid.

(SMS), which provides information on the level of risks and impacts of this phenomenon.[18] The system benefits from computer applications integrating:

- meteorological data, necessary for the development of climatic water balance (CWB),
- data from the digital soil-agricultural map, showing the spatial variation in water retention of different agronomic soil categories.

Two factors have been considered in delimiting the areas at risk of agricultural drought for individual crops within the drought monitoring system, namely: weather conditions and the vulnerability of soils to drought. This, in turn, makes it possible to identify areas vulnerable to economic losses due to drought affecting crops and agricultural production.[19]

2   The Polish concept of a Local water partnership

The concept of a 'local water partnership' was presented by the authors, Jarosław Gryz and Slawomir Gromadzki, on 22 May 2020 during an online seminar broadcast for invited stakeholders from the Ministry of Climate and Environment. The meeting was also attended by the Minister for Climate and Environment, Michał Kurtyka, and the Plenipotentiary of the Minister for Environment for Counteracting the Effects of Drought and the Shortage of Water Resources in the Environment, Lukasz Lange. The Plenipotentiary was provided with materials and assumptions, which were analysed with a view to their use in the drought special regulations under development. On 28 May 2020, the assumptions of the concept were handed over to the Deputy Director of the Department of Air Protection and Urban Policy of the Polish Ministry of Climate and Environment, Szymon Tumielewicz. In addition, the Minister for Climate and the Environment took patronage of a book by Jarosław Gryz, Slawomir Gromadzki entitled *Counteracting Drought. Water retention in Poland's crisis management system,* published by PWN S.A. in 2021. The book presents the assumptions for integrating water retention into the crisis management system, the basis for building 'local water partnerships' and the development of 'local water retention and conservation strategies'.[20]

---

18   The drought monitoring system (SMS) is developed and supervised by the Institute of Cultivation and Soil Science – National Research Institute (IUNG-PIB) in Puławy, on behalf of the Ministry of Agriculture and Rural Development in accordance with the provisions of the Act dated 7 July 2005 on insurance of agricultural crops and livestock (Journal of Laws 2005 No. 150 item 1249). The drought monitoring system has been developed on the basis of the Institute's long experience of modelling predictions of crop yields using a unique soil database characterising the variation in soil habitat and water retention across the country. Ibid.

19   System Monitoringu Suszy Rolniczej, Ministerstwo Rolnictwa i Rozwoju Wsi, accessed 21.12.2020, https://susza.iung.pulawy.pl/.

20   Jarosław Gryz, Sławomir Gromadzki, *Przeciwdziałanie suszy. Retencja wody w systemie zarządzania kryzysowego Polski* (Warszawa: Wydawnictwo Naukowe PWN, 2021), 12–95.

Presented during a seminar in 2020 at the Ministry of Climate and Environment, the partnership model was defined as a 'Water Partnership', consisting of a local government (municipal and district), a water company, a farmer or farmers (land and forest holders) with institutional and financial support from the regional and national level. It was stated that the Partnership will be tasked with:

- planning and implementation of municipal and district water retention strategies and plans.
- water retention management at local, regional, and national levels.
- system coordination of water abstraction and discharge at local, regional, state-wide level.
- building a comprehensive national critical infrastructure.[21]

During the second seminar, entitled "Counteracting drought", held on 6 May 2021 at the Ministry of Climate and Environment, involving Minister Michał Kurtyka, Plenipotentiary of the Minister of Climate and Environment for counteracting the effects of drought and water scarcity in the environment Lukasz Lange, Professor Jarosław Gryz, PhD Slawomir Gromadzki, a diagnosis of the current state of water management in Poland was undertaken.[22] Minister Michal Kurtyka asserted that we currently face a huge challenge in mitigating the effects of drought and water shortages. *"The periods of drought that have occurred in recent years have emphatically demonstrated that water resources in Poland are insufficient and that their consequences have been experienced not only by many branches of the economy, but also by the environment. This is why rational management of water resources is so important."* The 'My Water' programme, launched in 2020, is intended to help mitigate Poland's drought and reduce the risk of flooding by building rainwater retention facilities near homes. *"Last year, the 'My Water' programme was very popular, so we decided to continue it this year as well. The second round of the programme provides a further 100 million zlotys for Polish families to subsidise rainwater retention facilities."*[23] The seminar also included debates on 'Drought Prevention and Water Retention in Poland' and 'Drought Prevention and Water Retention in the New Green Deal of the European Union'.

During the seminar, the Minister for Climate and Environment, Michał Kurtyka, stated that the topic of water scarcity in Poland is a structural and systemic challenge that requires a long-term strategy making use in particular of the potential of nature, including wetlands or peat bogs and forest areas. These are natural, biological and environmental mechanisms that can be the best stores for water. In doing so, he emphasised how much retention potential there is in wetlands and peat bogs, where around 35 billion cubic metres of water can be stored nationwide, almost

---

21   Material submitted to the Plenipotentiary of the Minister of the Environment responsible for mitigating the effects of drought and the scarcity of water resources in the environment.

22   Seminarium „Przeciwdziałanie suszy" z udziałem ministra klimatu i środowiska, accessed 2.02.2022, https://www.facebook.com/MKiSGOVPL/videos/929728354236565/UzpfSTExODc3 MjU4NDgxMTczOTooMDk5NzA2NTEzMzgoOTcz/.

23   Ibid.

10 times more than is currently stored in artificial reservoirs. An additional 11 billion cubic metres of water could also be stored in forested areas. He also emphasised how, over the past decades, Poland has been affected by phenomena such as the deliberate drainage of land with great potential, as a consequence of which we are currently paying, as a country, a quantifiable price for such actions. Here, he drew attention in particular to the specific losses associated with agriculture in terms of crop yields, the impact of these activities on biodiversity and the shortfall in water resources necessary for the needs of the population.[24] Minister Kurtyka also identified how important public awareness of the challenges of water scarcity is today.

The Plenipotentiary of the Minister of Climate and Environment for Counteracting the Effects of Drought and Water Scarcity in the Environment presented the results of an analysis being carried out by the ministry on counteracting the effects of drought in 2020 by local governments. The data revealed that 52 municipalities out of 2,477 across the country are carrying out such tasks with a value of 11.4 million, representing a mere 2% of the total. In addition, municipalities in three provinces of Poland are not implementing such tasks at all, indicating a worrying phenomenon that points to a certain systemic gap and determines a need for action in this area. Professor Jarosław Gryz and Dr Slawomir Gromadzki indicated the need to involve as many partners as possible in drought mitigation efforts, including farmers and local governments. It was emphasised that retention alone cannot solve the effects of the drought faced by farmers, as water must be distributed to them. In addition, it was emphasised how important it is to verify the feasibility of implementing water retention within the state's crisis management system and the creation of so-called local water retention strategies, in which sites in given catchment areas with real retention potential can be designated.[25] The meeting at the Ministry of Climate and Environment was widely reported in the political, economic, and agricultural trade press.

3   Pilot project supporting the creation of local water partnerships

"Think globally, act locally" is the main thrust of a pilot programme for shaping water resources in agricultural areas under which Local Water Partnerships are to be established and coordinated by the Ministry of Agriculture and Rural Development, with the support of the Provincial Agricultural Advisory Centres. It is now recognised that "PGW Wody Polskie" is the administrator of almost all inland surface waters in the country. However, in addition to this area managed by PGW WP, there are also undeveloped watercourses, drainage ditches. *"These are located on private land – including individual farmers or water boards. This is where problems arise – with their proper management and maintenance. And, as a consequence, with*

---

24   Ibid.
25   Ibid.

drought".[26] With this it was made clear that the aim of the Local Water Partnerships is to create the right conditions and areas of cooperation between those who have an influence on water management in the designated areas.[27] By definition, the partners of the LAGs are local governments, representatives of the Polish Waters, water companies, farmers, individuals, community organisations and Agricultural Advisory Centres.

In order to extend the existing activities at central level, the Polish Ministry of Agriculture and Rural Development has worked on the creation of Local Water Partnerships (LWPs), which are designed to initiate and strengthen cooperation between all key partners (local social organisations, public entities, agricultural sector) for the management of water resources in agriculture and rural areas at local level. This project involves the application of a territorial approach to mitigate water management challenges in rural areas. In addition, it envisages the creation of 'an interface between relevant stakeholders with influence in this area, through the creation of local water partnerships (LPWs). As the area of operation of the LPW, the district was adopted as representative of the entities operating in its area (including local government units at district and municipal level)."[28] The Centre for Agricultural Advisory Services in Brwinów (CDR) is responsible for ensuring that water management tasks are conducted in a uniform manner throughout the whole country with the aid of provincial water coordinators employed at the Provincial Agricultural Advisory Centres, as well as appropriately prepared water specialists (training, instructions and materials). The first pilot partnerships were formed one in each province (less two provinces which formed two), for a total of eighteen pilot Local Water Partnerships to develop solutions to establish partnerships in each district. The experience gained will be used to implement appropriate solutions in the remaining districts.[29]

In assessing the solutions implemented involving the formation of local water partnerships under the auspices of the Ministry of Agriculture and Rural Development, it should be emphasised that their organisation takes place at district level. This is an unduly large administrative area which does not directly relate to the micro catchment as the primary functional area for retention planning. Therefore, associations of districts and municipalities territorially encompassing microcatchments would be more appropriate. It should also be noted that the district local authority:

---

26   R. Struzik, Tworzone są Lokalne Partnerstwa ds. Wody, PGW Wody Polskie, accessed 3.02.2022,
      https://www.terazsrodowisko.pl/aktualnosci/rolnictwo-niedobory-susza-Lokalne-Partnerstwa
      -ds-Wody-9093.html.

27   Ibid.

28   Lokalne partnerstwa ds. wody, Susza – Portal Gov.pl, accessed 3.02.2022, https://www.gov.pl
      /web/susza/lokalne-partnerstwa-wodne.

29   Ibid.

- performs statutory tasks in the field of water management (public water supply),
- develops local land use plans,
- may be involved in the work of water companies,
- and works directly with farmers and forest owners in the municipality.

In addition, the municipal government should design, initiate, supervise and develop local water retention strategies that will be part of the crisis response system from local (municipalities, districts) to national (provinces, state) levels.

In reviewing the solutions adopted, it should also be noted that there is a lack of a methodology for building partnerships and planning documents on the basis of which the partnership is to perform tasks in the field of water retention and counteracting the effects of drought. The concept of Local Water Partnerships (LPW) implemented by the MRiRW rightly identifies that it is essential to involve all water users whose decisions directly affect the quantity and quality of water in agriculture and rural areas as a common good of strategic importance. This asset should be treated as a legacy for future generations.

When considering the solutions adopted, it should be stated that rational water management in its broadest sense requires the capture of information on:
- water resources in a given area,
- meteorological forecasts for the current growing season,
- consumption estimates and analysis of water recovery rates,
- that which is essential for proper planning in this regard.

There is also a necessity for:
- the dissemination of good practices to optimise water consumption,
- the collection of water during periods of excess (downpours, snowmelt) for use during periods of shortage,
- identification of the needs of the whole population, with a particular focus on agricultural production, which is the basis of the country's food security,
- funding water management measures in rural areas from both Polish and European public funds, e.g., in the framework of the Green Deal strategy,
- strengthening coordination between stakeholders involved in the management of water resources in rural areas at regional and local level,
- effective implementation of public policies, alleviating access problems to water for agriculture and rural populations,
- establishing cooperation and joint action to promote sustainable water management.

The overall objective of the operation implemented by the Agricultural Advisory Centre entitled. "*Support for the creation of Local Water Partnerships*" was:
- establishing Poland's first network for cooperation between local communities and public institutions in the field of water management
- innovative support for LWP activities through the creation of an expert team, including scientific representatives, whose task would be to develop rules for the formation of LWPs,

- provision of training support,
- to produce a final report on the activities of the pilot group, identifying innovative solutions for the rational management of water in agriculture and
  rural areas,
- to initiate the activities of the coordinators of the emergent LWPs,
- preparation of water advisors of the 16 Provincial Agricultural Advisory Centres
  (WODR), whose task would be to initiate local water management activities.

The specific objectives of the project relate to practical forms of action. These are:

- "*Activation and integration of local communities through mutual awareness of
  each other's scopes of action and needs,*
- *diagnosis of the water resources management situation in terms of the needs of
  agriculture and rural residents,*
- *Developing common solutions for improving broader water management in agriculture and rural areas,*
- *Working out proposals for the future legal and organisational framework for the
  operation of LPW structures with the same tasks throughout the country,*
- Preparing a district area diagnostic report".[30]

Intended outcomes of the project include:

- "Activating local communities to take collective action on sustainable water
  management and the reduction of water pollution.
- *Diagnosis of the resources of the 16 counties and analysis of the main water problems in these areas. Gathering solution proposals from members of the newly
  formed LWPs, which will form the basis for recommendations and the development
  of a model for the functioning of future LWPs across the country.*
- *Establishing 16 pilot local water partnerships.*
- *Establishing a thematic network dedicated to water with the participation of LWPs
  to strengthen the transfer of knowledge and innovation in agriculture with the participation of agricultural advisory services, science, local governments, farmers and
  public entities.*
- *Dissemination of the concept of sustainable water management among farmers
  and rural residents.*
- Dissemination of good practices related to water use, collection and
  conservation".[31]

In the activities identified, the Agricultural Advisory Centre (CDR) was tasked with
coordinating the activities envisaged in the project and:

- developing an organisational and programme framework for the activities of
  the LWP,

---

30   Wsparcie dla tworzenia Lokalnych Partnerstw ds. Wody, accessed 3.02.2022, https://www.cdr
      .gov.pl/aktualnosci-instytucje/3367-wsparcie-dla-tworzenia-lokalnych-partnerstw-ds
      -wody-lpw.
31   Ibid.

- providing training for the staff of agricultural advisory units in the areas required by the LWP pilot,
- ongoing organisational and technical support for the running of the pilots,
- preparing a summary report based on the 16 sub-reports.

As part of the project, the Provincial Agricultural Advisory Centres (PACs) were tasked with:

- selecting an LWP coordinator from among the water advisors,
- selecting a district from their province where the pilot will be conducted,
- participating in training courses and meetings organised by the CDR on the development of LWPs,
- inviting stakeholders and individuals to cooperate within the LWP,
- organising LWP meetings, including possible field trips to showcase good practice,
- collecting materials, documents, information, and analysing questionnaires needed to prepare the report,
- developing the report.[32]

The project to support local water partnerships had a nationwide footprint. It was conducted in the second quarter of 2020 and in the first quarter of 2021. The following were identified as its beneficiaries: representatives of science, agricultural advisory units, representatives of the Polish Waters, local governments, the Ministry of Agriculture and Rural Development (MARD). Farmers and rural residents accessing water resources were identified as the end beneficiaries. In each province, the project supporting local water partnerships had its own characteristics and forms of organisation.

3.1 Pilot project in selected provinces of Poland

Local Water Partnerships (LWPs), according to the Ministry of Agriculture and Rural Development and the Agricultural Advisory Centre in Brwinów, working in cooperation with the Provincial Agricultural Advisory Centres, are intended to be a starting point for rational water management by creating a network of cooperation between partners: institutions, social organisations, individuals and stakeholders working to manage water resources in agriculture and rural areas at local level. The premise of these activities is to establish cooperation, with the aim of:

- diagnosing the state of water management,
- wypracowaniu metod racjonalnego gospodarowania wodą.

The pilot programme – Phase I of the project started in 2020 saw the establishment of at least one LWP in each district in every province. "The aim of the pilots undertaken was to integrate the local communities and to diagnose the water management situation in terms of agricultural needs and, as a consequence, to develop common solutions for water management in the region and to propose legal and

---

32   Ibid.

organisational standards for future LWP structures, as well as to develop a diagnostic report for the area of the relevant district."[33]

3.1.1 Małopolskie province

"In the Małopolska province, the project was implemented as part of the operation called 'Local Water Partnership (LPW)' provided for in the National Rural Network's Operational Plan 2020–2021 for SIR, co-financed by the European Union under the National Rural Network for Rural Development Programme 2014–2020."[34] The composition of the representative two county LWPs in the Małopolska province included various stakeholders (Table 4.1).

During the course of the project, during 2020, the following were realised:
- 6 meetings with partners (face-to-face and/or via teleconference) with a total of over 200 participants, addressing, among other things:
- LWP assumptions were discussed,
- mutual knowledge of each other's scopes of action was defined,
- exchange of existing experiences in water resource management
- thematic module on 'Principles of water use',
- thematic module on 'Functioning of water companies',
- thematic module entitled 'Dissemination of good water management and water efficiency practices in agriculture and rural areas',
- thematic module entitled 'Prospects and options for solutions (analysis of priority actions in the field of water, and support from EU or national funds for water-related investments)',
- preliminary versions of the final reports were presented.
- an action plan for the future was agreed.
- surveys were conducted on existing resources and activities related to water management (70 surveys among farmers in the Miechów district and 60 among farmers in the Proszowice district),
- information films: "Surface water erosion" and "Water companies in theory and practice",
- an information leaflet on 'The functioning of water companies'.
- information brochures: *"Principles of water use in the district of Miechow. Abstraction of surface and groundwater for irrigation." and "Principles of water use in the district of Proszowice. Water retention in agriculture".*[35]

The project culminated in the production of final reports:
- "Final Report of the Miechow District Local Water Partnership",
- "Final Report of the Proszowice District Local Water Partnership",

As part of the final conclusions, the need was identified for:

---

33   Ibid.
34   Water resource management – Local Water Partnerships (LPW), https://modr.pl/aktualnosc /zarzadzanie-zasobami-wody-lokalne-partnerstwa-ds-wody-lpw, accessed 3.03.2022.
35   Ibid.

TABLE 4.1   Partners of Local Water Partnerships

Miechów county and Proszowice county	Miechów county
1. Małopolski Urząd Wojewódzki	10. Starostwo Powiatowe w Miechowie
2. Urząd Marszałkowski Województwa Małopolskiego	11. Urząd Gminy i Miasta Miechów
3. Instytut Techniczno-Przyrodniczy Małopolski Ośrodek Badawczy w Krakowie	12. Urząd Gminy Charsznica
	13. Urząd Gminy Gołcza
	14. Urząd Gminy Kozłów
4. Uniwersytet Roliniczy im. Hugona Kołłątaja w Krakowie	15. Urząd Gminy Książ Wielki
5. Małopolski Oddział Regionalny Agencji Restrukturyzacji i Modernizacji Rolsnictwa w Krakowie	16. Urząd Gminy Racławice
	17. Urząd Gminy Słaboszów
	18. Regionalny Związek Spółek Wodnych w Jędrzejowie
6. Regionalny Zarząd Gospodarki Wodnej w Krakowie	19. Zakład Wodociągów i Kanalizacji w Miechowie Sp. z.o.o.
7. Regionalna Dyrekcja Ochrony Środowiska w Krakowie	20. Zespół Parków Krajobrazowych Województwa Małopolskiego
8. Wojewódzki Fundusz Ochrony Środowiska i Gospodarki Wodnej w Krakowie	21. Nadleśnictwo Miechów
	22. Powiatowy Zespół Doradztwa Rolniczego w Miechowie
9. Małopolska Izba Rolnicza	23. Osoba fizyczna z gminy Książ Wielki
	Proszowice County
	10. Starostwo Powiatowe w Proszowicach
	11. Urząd Gminy i Miasta Koszyce
	12. Urząd Gminy i Miasta Nowe Brzesko
	13. Urząd Gminy i Miasta Proszowice
	14. Urząd Gminy Koniusza
	15. Urząd Gminy Pałecznica
	16. Urząd Gminy Radziemice
	17. Wodociągi Proszowickie Spółka z.o.o.
	18. Powiatowy Zespół Doradztwa Rolniczego w Proszowicach

SOURCE: WATER RESOURCE MANAGEMENT – LOCAL WATER PARTNERSHIPS (LPW),
HTTPS://MODR.PL/AKTUALNOSC/ZARZADZANIE-ZASOBAMI-WODY-LOKALNE-PARTNERSTWA
-DS-WODY-LPW, ACCESSED 3.03.2022

- *"Solutions for the lack of a sewage network;*
- *review and assessment of the condition of existing basic and secondary drainage systems.*
- *modernisation of the existing drainage network (controlled drainage).*
- *increasing soil and landscape water retention.*
- *concepts for the local restoration of rivers and streams.*
- maintenance/establishment of water associations to deal with the maintenance and upkeep of drainage".[36]

The reports also indicated: the need to involve partners, the need for legal solutions for the LWP structures to be created, securing sources of funding for water improvement and shaping projects.

In 2021, Phase II of the project was launched, with the aim of successively establishing LWPs in further districts in each province, and ultimately establishing LWPs in every district in the country. W województwie małopolskim projekt realizowany jest w ramach operacji pt. *"Local Water Partnership (LWP) in Małopolska" provided for in the National Rural Network's Operational Plan 2020–2021 in the scope of SIR, co-financed by the European Union under the National Rural Network of the Rural Development Programme 2014–2020.* It was assumed that the work would proceed in groups of districts reflecting similarities in geographical and economic/social conditions, agro technology, groundwater and surface water resources. It was also accepted that long-term strategic plans would be created with a list of investments to improve water management.[37]

3.1.2   Lubuskie province

Since 2020, the Lubuskie Agricultural Advisory Centre in Kalsko has been delivering a task commissioned by the Agricultural Advisory Centre in Brwinów and the Ministry of Agriculture and Rural Development to create a Local Water Partnership.[38] The project covers five districts of the Lubuskie Province, namely Świebodzin, Gorzów, Międzyrzecz, Zielona Góra and Krosno. In 2021, a series of meetings were held in each of these districts Partners in the project included local authorities, forestry departments, the marshal's office and the chamber of agriculture. It was noted that Local Water Partnerships could solve many water and water management problems at a local level in the future.[39]

---

36   Ibid.
37   Zarządzanie zasobami wody – Lokalne Partnerstwa ds. Wody (LPW), accesed 20.09.2022, https://modr.pl/aktualnosc/zarzadzanie-zasobami-wody-lokalne-partnerstwa-ds-wody-lpw.
38   The operation is co-financed by the European Union, under Scheme II of the Technical Assistance "National Rural Network" of the Rural Development Programme 2014–2020 the Managing Authority of the Rural Development Programme 2014–2020 is the Ministry of Agriculture and Rural Development.
39   Local water partnership – second meeting, accessed 3.02.2022, lodr.pl.

3.1.3   Kujawsko-Pomorskie province

The Kujawsko-Pomorski Agricultural Advisory Centre in Minikowo, on behalf of the Ministry of Agriculture and Rural Development, started the task of establishing local water partnerships on 25 June 2020. This took place in the two districts of Sepolenski and Nakiel. In order to realise the idea of LWP, i.e. the integration of partners interested in counteracting the effects of drought and undertaking activities aimed at building small and medium-sized retention facilities, the following were invited to cooperate: district and municipality self-governments, local farmers' organisations (the Kujawsko-Pomorska Chamber of Agriculture), the Provincial Office in Bydgoszcz, the Marshal's Office in Toruń, the State Forests, the Krajeńskie Landscape Park, water companies, Local Action Groups, Local Fishery Groups, and the State Water Management Company Wody Polskie. Dr Ryszard Kamiński, Deputy Minister of Agriculture and Rural Development (initiator of the LWP activities at the level of the Ministry of Agriculture and Rural Development), also participated in the LWP-related work and presented the assumptions for new legal regulations related to national water management. According to Dr Ryszard Zarudzki, Director of KPODR in Minikowo, the work undertaken is "one of the most important priorities planned for implementation in the next few years. Water partnerships are not only a consideration for water management, they also challenge local authorities, representatives of state forests, landscape parks and, of course, farmers to work together to make the joint investments needed to retain water in an area when there is an overabundance and then, in times of scarcity, to mobilise stored reserves for farming, forestry, the local community or for animals, not just livestock. Therefore, among other things, the aim of the proposed regulations is to introduce a package of solutions that will facilitate the retention of water and improve the availability of water resources in order to mitigate the negative effects brought about by increasingly prolonged periods of drought. These undertakings should be aimed at preserving, creating and restoring water retention systems or preventing the reduction of surface and groundwater levels".[40] The aim of the KPODR in Minikowo is to "support the creation of Local Water Partnerships", which will assume "the creation of a network of cooperation between local society in the field of water management, the development of a model for the functioning of future LWPs throughout the province, and the development of innovations in the field of water management in agriculture and the countryside".[41]

The main results of the working group's work under the auspices of the Kujawsko-Pomorskie Agricultural Advisory Centre in Minikowo were the adoption

---

40   M. Kołacz, Wojewódzki Koordynator Lokalnych Partnerstw Wodnych – KPODR w Minikowie, Lokalne Partnerstwa Wodne (LPW) – w powiatach sępoleńskim i nakielskim – Sieć Innowacji w Rolnictwie i na Obszarach Wiejskich (kpodr.pl).

41   Michał Kołacz, Wojewódzki Koordynator Lokalnych Partnerstw Wodnych – KPODR w Minikowie, Lokalne Partnerstwa Wodne (LPW) – w powiatach sępoleńskim i nakielskim – Sieć Innowacji w Rolnictwie i na Obszarach Wiejskich, accessed 2.02.2022, kpodr.pl.

of arrangements for the operation of the partnership in the district and the collection of data and information for drawing up the report. The agreed principles and objectives of the Local Water Partnership are:

- *"Working together for rational water management by creating mechanisms to ensure that partners participate in decision-making and investment activities.*
- *Establishing a system within the district for the flow of information, consultation and coordination between all stakeholders involved in investment and regeneration activities in the field of water management.*
- *Raising awareness of rational water management among residents and businesses in the district.*
- *Triggering various social initiatives for rational water management by promoting this issue.*
- *Building good relations between water stakeholders, including raising the profile of Water Companies as an important factor in shaping water relations.*
- *Creating instruments to assist partners and farmers in the development of planning documents, analytical documents and financial proposals related to water investment objectives of the operation to support the creation of Local Water Partnerships".*[42]

On 28 January 2021, the Kujawsko-Pomorski Agricultural Advisory Centre in Minikowo also hosted an online conference entitled: "Local Water Partnerships as an Opportunity for Rational Water Management in the Kuyavian-Pomeranian Province", during which a summary of the pilot programme in the districts of Sępolno and Nakło was presented.[43]

3.1.4   Podlaskie province

In the province, the implementation of projects relating to Local Water Partnerships was undertaken by the Podlaskie Agricultural Advisory Centre in Szepietów. The venue for the pilot partnership project became the Grajewski district. On 27 July 2020, the first LWP meeting was held in Grajewo. The partnership comprised the following stakeholders: District Starosty in Grajewo, Grajewo Municipality, Radziłów Municipality, Town and Municipality of Szczuczyn, the Regional Water Catchment Board in Białystok, Biebrza National Park, water companies, Polish Water, state forests, national parks, chambers of agriculture and farmers. The involvement and water management competences of the different stakeholders and partners of LPW were identified. The meeting concluded with the signing of a declaration of cooperation within the framework of the Grajewo District Local Water Partnership.[44]

---

42   Ibid.

43   Kujawsko-Pomorski Ośrodek Doradztwa Rolniczego w Minikowie, accessed 3.02.2022, https://kpodr.clickmeeting.com/webinar-recording/fLpW30a86.

44   LPW, czyli Lokalne Partnerstwo Wodne – PODR w Szepietowie, accessed 3.02.2022, LPW, czyli Lokalne Partnerstwo Wodne – PODR w Szepietowie.

3.1.5   Lublin province

In the Lubelskie Province, a meeting on the formation of local water partnerships was held on 7 September 2021 at the initiative of the LAG 'Lepsza Przyszłość Ziemi Rycka'. During its course, the following was accomplished:
– identification of water efficiency problems,
– preparation of a list of investment tasks to improve the water balance,
– identification of potential funding sources.

The meeting stressed that "the drought and water shortage in an era of ongoing climate change shows how important and necessary comprehensive and sustainable water management is. The long-standing neglect of water management, drainage and retention in the country is very high. The quality of life of today's and future generations living in the Polish countryside will depend on our actions in water management".[45]

3.1.6   Lódz province

The Łódzki Agricultural Advisory Centre in Bratoszewice held a series of meetings entitled 'Local Partnerships for Water 2021', in the districts of Łowicz, Brzeziński, Sieradz, Łęczyca, Poddębice, Tomaszów, Opoczno and Wieruszów. They were part of the KSOW Operational Plan 2020–2021, co-financed by the European Union under Scheme II of the Technical Assistance "National Rural Network" of the Rural Development Programme 2014–2020. *The main theme of the meetings was 'Rational water management on agricultural land in times of climate change. Vision for the functioning of the Local Water Partnership in the districts of the Łódź Province".* In doing so, the following issues were addressed:
– "The scientific basis of water management in agricultural areas,
– *landscape and riverbed retention – a strategic area in the process of adapting agriculture to climate change and addressing water deficits in agricultural productive space.*
– *good practice – improving water quality,*
– *the condition of water resources in the district".*[46]

These issues were raised as part of the Local Water Partnerships 2021 project. Their main objective was to adopt an innovative approach to drought mitigation activities in rural areas of the Łódzkie Province by creating groups of people and stakeholders who work together to activate and integrate rural residents as well as those responsible for water management in the area.

---

45   Lokalne Partnerstwa do Spraw Wody 2021 | Łódzki Ośrodek Doradztwa Rolniczego w Bratoszewicach, accessed 3.02.2022, Lokalne Partnerstwa do Spraw Wody 2021, lodr-bratoszewice.pl.
46   Ibid.

3.1.7 Opolskie province

The Opolski Agricultural Advisory Services Centre in Łosiów started organising 'Thematic meetings on the establishment of local water partnerships (LWP)' in 2020. This took place within the framework of the 2020–2021 biennial operational plan for the Network for Innovation in Agriculture and Rural Areas. The main topic of the meetings was water management and the development of guidelines for the creation of a Local Water Partnership in the Krapkowice district. This activity culminated in the drafting of:

- "a report on the formation and work of the Krapkowice District Local Water Partnership",
- informational material 'Good water management and water efficiency practices in agriculture and rural areas'.

The subsequent step was to organise thematic meetings in an online format in the districts of Strzelce, Olesko, Głubczycki, Opolski, Kędzierzynsko-Kozielski, Namysłowski, Brzeski, Prudnicki, Nyski and Kluczbork. The aim of the meetings was to activate and integrate local communities in the framework of local water partnerships. The themed meetings focused on:

- the concept and role of local water partnerships,
- raising participants' awareness of drought and how to minimise its impact,
- water requirements of agricultural production,
- legal standards for water law,
- the operation of water companies,
- water management diagnosis as a fundamental basis for the actions of future partnerships.
- knowledge transfer across local water partnerships,
- raising awareness of good water management practices,
- the proper functioning and support of water companies,
- funding for water investments,
- water use regulations and water permits.[47]

3.1.8 Silesia province

In the Silesian Province, an operational plan for 2020–2021 called 'Establishment of Local Water Partnerships in the Silesian Province' was undertaken by authorities as part of the National Rural Network Action Plan 2014–2020. The Silesian Agricultural Advisory Centre in Częstochowa, which unites farmers, local government representatives, water companies and water users interested in water management, was responsible for implementing the project. The project responds to the plans of the Ministry of Agriculture and Rural Development and the Agricultural Advisory Centre in Brwinów to establish Local Water Partnerships in every district of every

---

47 Tworzenie Lokalnych Partnerstw ds. Wody (LPW) w Województwie Opolskim | Opolski Ośrodek Doradztwa Rolniczego, accessed 3.02.2022, oodr.pl.

province. The project focused on potential opportunities to improve water availability, address drought, flood risks, protect water ecosystems from pollution, and the use of surface and groundwater.[48]

3.1.9   Świętokrzyskie province

In the Świętokrzyskie Province, a pilot programme on the creation of a Local Water Partnership was realised in the Konecki District as part of the Network for Innovation in Agriculture and Rural Areas (SIR), entitled 'Local Water Partnerships. "Establishing links between stakeholders interested in the creation of a Local Water Partnership in the district of Koneck". – Operational Plan 2020–2021. The operation established a cooperation platform between community partners, namely: The local community and public administration in the field of water management, as well as for agriculture. The Local Water Partnership formed as part of the pilot met four times and the end result was a final report entitled: "Local Water Partnership". "Analysis of solutions for the operation of the Local Water Partnership (LWP) – Konecki district".[49]

3.1.10   Warmia and Mazury province

Local Water Partnerships in Warmia and Mazury began to be formed as part of Phase I in 2020. The Warmia and Mazury Agricultural Advisory Centre in Olsztyn was responsible for implementing the project. The project was launched in the district of Braniewo and expanded to a further nine districts in 2021. According to the assumptions, the aim of the Local Water Partnerships is:
– identification of the region's water resources, and an estimate of requirements for production and domestic use,
– raising public awareness,
– activating farmers,
– rational water management in agriculture and rural areas.
– seeking innovative solutions and methods for water management,
– establishing water efficiency programmes.
It was found that each district has different characteristics, in terms of reported problems and investment expectations, which is important from the point of view of investment and planning activities. The Warmia and Mazury Agricultural Advisory Centre also planned to conduct remote sensing surveys on farms using drones to determine:
– wetlands
– areas experiencing water stress,
– the direction of water run-off,

---

48   Local partnership for water in your district – SIR, accessed 7.02.2022, odr.net.pl.
49   Lokalne Partnerstwo ds. Wody w powiecie koneckim – podsumowanie projektu | Świętokrzyski Ośrodek Doradztwa Rolniczego w Modliszewicach, accessed 7.02.2022, sodr.pl.

- drainage channel capacity,
- topsoil moisture content,
- areas and degree of pollution of water bodies.[50]

3.2 *National level water retention cooperation*

One of the key elements that can ensure the success of water retention projects is cooperation between the farming community (National Council of Chambers of Agriculture) and Polish Water, as confirmed by, among others, PGW WP President Przemysław Daca. PGW Wody Polskie acknowledges that the standard methods used so far to reduce runoff for drought mitigation are insufficient. The closure of damming installations during the April-May period, after the passage of snow-melt water is increasingly proving inadequate. *"Already in February, the State Water Company Wody Polskie took the decision to close all kinds of damming installations on rivers – of course, taking into account the current hydro-meteorological conditions."*[51] That is because there was no winter snow in 2020 /2021, neither did meltwater pose a flood risk. Moreover, the onset of drought was already recorded in early spring. PGW Wody Polskie "when considering non-standard measures"[52] came to the conclusion that something more needed to be proposed to counter the drought. A decision was made to implement the riverbed retention concept, which stipulated that each of the 50 Polish river basin districts would identify the watercourses in their area of greatest importance for agricultural production. On this basis, 627 projects were identified for construction / modernisation of water facilities, such as weirs, barrages and other hydrotechnical equipment. Implementation of the programme is expected to increase retained water resources by approximately 32 million cubic metres of water. The programme has a budget of 157 million zloty with a three-year expenditure timeframe. The implementation of the programme involves working with local authorities to jointly fund investment projects. PGW Wody Polskie admits, however, that the riverbed retention programme "will only allow the retention of 1% of the water in Poland".[53] This will be despite the relatively low cost and rapid outcomes. According to estimates, about 300,000 hectares of fields will be directly and indirectly irrigated in this way.[54]

Another positive example from the Wielkopolskie Province is the municipality of Rogozno, whose authorities have established cooperation with the Regional Water Management Authority in Poznan and the State Forests. "Forest areas adjacent to the Vlna and the Little Vlna will be designated, areas where the water level

---

50 Warmia and Mazury Agricultural Advisory Centre in Olsztyn, accessed 7.02.2022, wmodr.pl.
51 Tworzone są Lokalne Partnerstwa ds. Wody PGW Wody Polskie, accessed 7.02.2022, teraz-srodowisko.pl.
52 Ibid.
53 Ibid.
54 The municipality of Drelów, where more than 500 ha of land is irrigated thanks to water storage, is cited as a good example of the effectiveness of trough retention. Ibid.

will be temporarily raised by, with the construction of levees, (using natural materials, e.g., stone, wood). This will have the effect of slowing down outflows in the first instance and preserving inviolable flows during periods of depletion. On the other hand, during times of elevated water levels, it will be possible to temporarily flood forest areas, raise groundwater levels and distribute excess water to the surrounding forests, thanks to Partnerships, it is possible to share this work.

In March 2021, the issue of Local Water Partnerships activities was addressed by the Senate Environment and Agriculture and Rural Development Committees in a joint meeting. The Deputy Minister of the Ministry of Agriculture and Rural Development, A. Gembicka "stressed that the Local Water Partnership, with the district as its centre of action, can include local governments at all levels, the Polish Water Authority, water companies, working farmers and a range of other stakeholders. In the first phase, conducted in 2020, 18 Partnerships were operating within the framework of the pilot, aiming to diagnose the current situation and develop investment and legal/organisational proposals to be established in each district". At the provincial level, the provincial agricultural advisory centres are responsible for coordinating the activities, and the overall system is coordinated by the Agricultural Advisory Centre in Brwinów. In 2021, the Agricultural Advisory Centre in Brwinów identified plans to establish Partnerships in 165 districts. According to Ryszard Zarudzki, director of the Kujawsko-Pomorskie Agricultural Advisory Centre, "the issue of water for rural areas was included in the Strategy for Responsible Development and the measures currently being taken are actually a consequence of the provisions of the strategy for responsible development".[55]

4   Conclusions

Existing organisational solutions in Poland are adapted to the implementation of measures combining flood and drought prevention within the framework of the water resources protection strategy, its retention in catchments as well as in river basins. The Drought Plan adopted in 2021 lays the groundwork for the activity of public institutions, social organisations, farmers and other stakeholders. Drought Plan lays the groundwork for the activity of public institutions, social organisations, farmers and other actors. The optimal formula for the selection of activities, as well as the means of achieving them over time, is derived from the implementation of the water protection concept using the LIDER method and the bottom-up, partnership approach to rural development implemented by the Local Action Group (LAG). At the same time, there is a need to combine them in a national strategy for dealing with the effects of drought and water shortages in Poland. The issue concerns:

---

55   W Senacie Lokalnych Partnerstwach ds. Wody, Dziennik Warto Wiedzieć, accessed 8.02.2022, wartowiedziec.pl.

1. Linking Local Development Strategy (LSR) through the local rural or urban community centred in it, the implementation of joint, innovative projects, combining human, natural, cultural, historical, and other resources in the projection of activities of state institutions and local government bodies. This in Polish practice means combining the activities of: Agricultural advisory centres, provincial offices, marshal offices, district offices, the State Water Management Authority Wody Polskie (comprising the National Water Management Authority, eleven regional water management boards, fifty catchment management boards, three hundred and thirty water superintendents), the State Forestry Authority Lasy Państwowe (State Forests), municipal governments, water companies, farmers' associations and forest owners, and other (social, economic) stakeholders.

2. The national implementation of Local Water Partnerships requires not only the debate that took place in 2019–2020, but above all the inclusion of stakeholders in the form of farmers, other water resource users. Meetings of local authorities, municipalities with farmers, and forest owners should be the start of bottom-up water retention planning. In the current described formula of activities, there is a lack of mayors, village leaders, farmers, and other stakeholders and at the same time beneficiaries of the Local Development Strategy for the protection of water resources.

3. The investments identified in the Drought Plan are scattered and have very little impact on the problem of drought and flood prevention. Central planning, an offshoot of a bygone political system, will not work with the changing investment needs of farmers, and other stakeholders, which are dictated by climate change. The model described above needs to be reconfigured in Poland, adopting a bottom-up formula of initiatives undertaken by Local Action Groups using the LIDER method as a benchmark for all European Union countries.

4. The solutions identified as desirable should be tested within the Local Action Groups using the LIDER method in catchment areas and river basins and then subjected to financial costings in terms of the necessary funding. Given the scale of the project – covering the whole of Poland – the Drought Plan should be aligned in time with Local Action Group investments. Taking into account the time, the scope of investments, the financial means to implement them, this plan should follow a seven-year investment period aligned with budgeting within the European Union so that it becomes embedded in the New Green Deal.

CHAPTER 5

A universal European model for local water partnerships

The universal European model of local water partnership includes several elements critical to the success of the implementation of the idea. The first is its concept of implementing the objectives of the European Union's New Green Deal. The second, a formula that allows it to be implemented using existing EU instruments as is the case in individual Member States. Thirdly, to identify how to use existing solutions to implement local water partnerships within the European Union countries, involving their national and local institutions and stakeholders.

1   Approach to model building

Preliminary research where included in the paper "Model of a European 'Local Water Partnership'. Delivering the European Green Deal". They where suplemented by thoughts delivered in "Tackling drought. Water retention in Poland's crisis management system". All together was compared with the result of the analysis discussed so far were used as input data for the modelling method. As a result of which, the assumptions of a new system model of water resources management based on 1 "Local Water Partnership", 2. "Local water retention and protection strategy". The model was also conceptualised based on the empirical findings of Slawomir Gromadzki, based on:

- his own experience as a farmer running an organic farm located in north-eastern Poland in areas experiencing drought.
- internship and placement (6 months) as part of the European project 'Harmonising water related graduate education' Fellowship – EACEA Erasmus + Water Harmony Project in which two specialist courses were completed:
  - Coursework: Water Resources Management and Treatment Technologies, Norwegian University of Life Sciences, As.
  - Specialised course: Water management in cold climate, University Centre in Svalbard.
  - Experiences as an Expert / Member of the Focus Group 46: Water: Nature-Based Solutions for water management under climate change, The agricultural European Innovation Partnership (EIP-AGRI), European Commission,
- Slawomir Gromadzki own research in developing the assumptions of the European water conservation retention model contained in the work Model of a European "Local Water Partnership". Delivering the objectives of the European Green Deal.

The cognitive scope of the model, its form defined within the framework of the security crisis management of the state, and international organisation, defined:

- Research, by Jarosław Gryz, and Slawomir Gromadzki, which developed a national model for a system-wide solution to the problem of water retention in Poland by linking retention to a nationwide crisis management system.
- Sławomir Gromadzki experience as head of the Municipal Response Team in the Municipality of Mały Płock and as initiator, co-founder and first chairman of the Local Action Group "Kraina Mlekiem Płynąca".
- Jarosław Gryz experience of working in the security establishment of Poland, the North Atlantic Alliance, focused on national and international strategic security management.[1]

In relation to this approach, the basis for building the model was: 1 Adopting the discussed LEADER method as a tool for planning and implementing local development strategies, i.e., "Local Water Partnership". 2. Identification of assumptions for a "Local Water Retention and Protection Strategy". 3. Defining a framework for the implementation of the above-mentioned solutions in the practice of a European Union Member State as well as the organisation itself.

2   Normative basis for the model

The paper reviews the European Union's leading systems approaches to water resource management. In doing so, reference was made to the following policies and legislation:

- Communication from the Commission to the European Parliament, the European Council, the Council, the Economic and Social Committee and the Committee of the Regions The European Green Deal.
- Directive 2000/60/EC of the European Parliament and of the Council dated 23 October 2000 establishing a framework for Community action in the field of water policy.
- Commission Directive 2014/101/EU dated 30 October 2014 amending Directive 2000/60/EC of the European Parliament and of the Council establishing a framework for Community action in the field of water policy.

These documents were used to verify the solutions in terms of the provisions analysed, and their interpretation in terms of the model presented.

---

1 Jarosław Gryz, activities in years:
   2019–2021 The Science for Peace and Security (SPS) Programme, participant in working panel.
   2016 Strategic Defence Review, Ministry of Defence, working panel participant.
   2010 Strategic National Security Review, Office of National Security, working panel participant.
   2008–2009 Multiple Futures Project, NATO's Allied Command Transformation, expert.
   2007 National Security Strategy, National Defence Academy, participant in working panel.

2.1 *Water in the European Green Deal strategy*

An analysis of the 'European Green Deal' from the perspective of determining the importance of water in shaping the strategy's objectives contains two elements relating to this issue. In Chapter 2 'Transforming the EU economy for a sustainable future', subsection 2.1 'Developing a set of policies that will bring about deep trans-formation' is included. Within it, key references to the management and protection of water resources are identified for each of the policies adopted. Political strategies are defined here based on *2.1.3. Mobilising the industrial sector in support of a clean, closed-loop economy.* In relation to these records, it was found that "Approximately half of total greenhouse gas emissions and more than 90 per cent of biodiversity loss and water deficit are caused by resource extraction and the processing of raw materials, fuels and food." Having identified this problem, it was shown that "Digital technologies are key to achieving the Green Deal's sustainability goals across a wide range of sectors." In addition, "Digitisation offers new opportunities for remote monitoring of water and air pollution and for monitoring and optimising the use of energy and natural resources." The above interpretation sets out the preferred course of action.

In the provisions of point 2.1.6. From field to table: creating a fair, healthy and environmentally friendly food system, it was stated: "European food is renowned for its safety, nutritional richness and high quality. It should also become a global standard for sustainability. The transition to more sustainable systems has already begun, but with current food production methods, feeding the world's rapidly growing population is still a challenge. Food production continues to pollute the air, water and soil, contribute to biodiversity loss and climate change and consume vast amounts of natural resources, while wasting a large proportion of the food pro-duced. Poor quality food contributes to obesity and diseases such as cancer." The identification of this problem made it possible to identify measures to counter it. "The Commission will work with Member States and stakeholders to ensure that national strategic plans in this area fully reflect the ambitions of the Green Deal and the farm-to-table strategy from the outset." "Plans should lead to sustainable practices such as precision farming, organic farming, agroecology, agroforestry and stricter animal welfare standards. Shifting the focus from compliance to efficiency, measures such as eco programmes should reward farmers for better delivery of environmental and climate goals, including soil carbon management and storage, and better nutrient management to improve water quality and reduce emissions. The Commission will work with Member States to increase the potential of sustain-able seafood as a source of low-carbon food."

The provisions of the document, section 2.1.7. Protecting and restoring ecosys-tems and biodiversity, they state "Ecosystems perform essential services, providing food, fresh water, clean air and shelter. They help mitigate the risk of natural disas-ters, reduce the incidence of pests and diseases and contribute to climate regulation. However, the EU is failing to meet some of its most important environmental tar-gets for 2020, such as the Aichi targets adopted under the Convention on Biological

Diversity. The EU and its global partners must halt the loss of biodiversity." The remedy for the described state of affairs is to be: "A sustainable blue economy has a central role to play in reducing different types of land resource needs in the EU and tackling climate change. There is growing recognition of the role that the oceans play in climate change mitigation and adaptation. The sector can contribute to the green transition by improving the use of water and marine resources". The indicated provisions define the nature of the activities contained in section 2.1.8. *Zero emissions for a non-toxic environment.* "In order to protect citizens and ecosystems in Europe, the EU needs to better monitor, report, prevent and remedy air, water, soil and consumer product pollution. To this end, the EU and its Member States will need to examine all policies and regulations in a more systematic way. With a view to addressing these related issues, the Commission will adopt an action plan for the elimination of air, water and soil pollution in 2021." Of particular relevance in section 2.1.7 is the statement, "The natural functions of surface and groundwater must be restored. This requires the protection and restoration of biodiversity in rivers, lakes, wetlands and estuaries. The restoration of these functions is also needed to prevent and reduce losses arising from flooding. The introduction of a farm-to-table strategy will reduce environmental pollution associated with excess nutrients. In addition, the Commission will propose measures to address pollution from urban runoff and new, particularly harmful pollutants such as microplastics and chemicals, including medicinal products. The cumulative effects of different pollutants also need to be addressed."

The excerpts cited from the 'European Green Deal' in terms of defining the importance of water revolve around protecting this resource, enhancing its quality and integrating it into the ecological economy of capitalism as a platform for implementing sustainable development. From this perspective, the implementation of the concept of a universal European model of local water partnerships is de facto part of the ecological transformation of European Union societies. Furthermore, it implements in practice the aforementioned objectives of reducing carbon emissions, soil pollution, preserving and enhancing biodiversity. It also fulfils the demand for the implementation of green capitalism in the pursuit of a farm-to-table strategy.

2.2   *Water directive*

It will only be possible to achieve the water protection objectives set out in the policy documents by adopting an innovative, EU-wide coherent delivery method for the European Green Deal. In order to achieve this, it will be necessary to amend Directive 2000/60/EC of the European Parliament and of the Council dated 23 October 2000 establishing a framework for Community action in the field of water policy,[2] in particular "Annex VII River Basin Management Plans" accepting

---

2   Directive 2000/60/EC of the European Parliament and of the Council of 23 October 2000 establishing a framework for Community action in the field of water policy, Official journal of the European Communities 22.12.2000.

the assumptions of the proposed model. This includes restoring degraded ecosystems across Europe by scaling up organic farming and biodiversity-rich landscape elements on agricultural land, halting and reversing the decline of pollinators, planting 3 billion trees by 2030, and addressing natural capital issues.[3]

2.2.1 Theoretical normative assumptions of the model

These include the proposed model of a local water partnership involving a local action group. The matter concerns the definition of the methods and directions of action by public partners (institutions of the state, regional composite and non-complex administrations), water companies (integrating entities of local self-government, private owners of agricultural land, forests, waters), social entities (local communities, NGOs) within the framework of crisis management of a state in the European Union. The normative interpretation should include:

– an integrated crisis management system formula covering the whole country for (macro and micro) surface water catchments,
– protection of water resources (surface, underground),
– the provision of water for agriculture using small and medium-sized retention in microcatchments,
– prevention of droughts, floods, contamination of watercourses, and fire protection,
– parameterisation of information from non-urbanised and urbanised areas covering micro-catchments for national security management,
– identifying practical, accessible means, tools, methods and combining them into a synergistic, coherent whole in the national security management system,
– identify directions for organisational and normative change in European Union countries favouring the use of micro retention in the utilisation and protection of water resources using new techniques and technologies (e.g., situational awareness, Internet of Things) that are available and being implemented gradually.

2.2.2 Theoretical organizational assumptions of the model

The creation of an interpretation of water retention in the prevention of droughts, floods, creating conditions for the protection and enhancement of water resources, their use in agriculture and forestry, protection of the environment, laying the foundations of a universal water supply system for agricultural production requires the fulfilment of several criteria, related actions.

Firstly, verification of knowledge on how to integrate water retention into the system of action of European Union countries for the protection of water resources (surface and groundwater), and the prevention of droughts and floods.

Secondly, the implementation of the idea that local water retention planning and management will be the responsibility of local water partnerships consisting of

---

3 Communication from the Commission to the European Parliament, the Council, the European Economic and Social Committee and the Committee of the Regions a European strategy for data, European Commission, Brussels, 19.2.2020, COM(2020) 66 final.

local government (municipal and district, water company, farmers/forest holders), social organisations, with organisational and financial support from the regional level, the state, and the European Union.

Thirdly, activities undertaken by local water partnerships consisting of local government (municipal and county government, water company, farmers/forest owners), community organisations will be included in a local water retention and protection strategy consistent with the spatial development plan, socio-economic development strategy (municipality), Local Development Strategy (Local Action Group/Local Fishery Group), crisis management plan for the respective area of an EU country.

Fourthly, the local water retention strategy for countering droughts, floods, creating conditions for protecting and increasing water resources, its use in agriculture and forestry, protecting the environment, building the basis of a universal water supply system for agricultural production should be configured in three synergistically linked bottom-up planning dimensions:

1. Flood protection.
2. Countering the effects of drought.
3. Protection against water contamination/pollution.

As a result, foundations will be laid within the framework of the European Union for the implementation of a universal agricultural water supply system for agricultural production in line with the idea of the New Green Deal.

2.3   *Proposed local water partnership model*

The study assumes that a 'Local Water Partnership' is an organisational form of grouping/association/union of public, social, private partners to implement a Local Water Retention and Protection Strategy' in relation to the stated objectives of the European Green Deal strategy, including:

1) Development and implementation of a local water retention and protection strategy based on a functional area such as a micro catchment possibly a small catchment, taking into account hydrological, climatic, geological, urban, technical, agricultural, forestry and other relevant conditions.

2) Development and implementation of a local water retention and protection plan.

3) Development and implementation of a local plan for the supply of water to agriculture and industry for production purposes (agricultural, industrial), in line with the 'farm to table' policy.

4) Coordination of water retention construction activities.

5) Coordination of water protection measures.[4]

---

4   Council Directive of 12 December 1991 concerning the protection of waters against pollution caused by nitrates from agricultural sources (91/676/EEC).

6) Coordination of measures to protect underground water resources.[5]
7) Water quality enhancement.[6]
8) Implementation of expert tasks (studies, plans, maps, feasibility studies, projects).
9) Implementation of investment tasks.
10) Implementation of training and education activities, and cooperation/participation in the flood prevention system.[7]

Each entity (legal entity and individual) operating within a 'Water Partnership' has the opportunity to act independently, according to its domain and vocation (assigned social, public, economic function). By participating in a "Local Water Partnership" being an integral part of it and committing its human, organisational and material resources to its operation, it will be able to participate in the achievement of the overall objective and its own specific objective. The assumption is that this will result in the partnership – as the sum of the actions of many diverse actors – acquiring properties that none of its constituent parts have. Therefore, the organisational and functional model for the "Water Partnership" are the Local Action Groups that have been operating in Poland since 2007, with attributes derived from the LEADER method, which they fully implement:

- Bottom-up initiatives (broad community participation in strategy development and implementation).
- Territoriality of initiatives (local development strategy prepared for a defined, geographically coherent area).
- Integration (linking of different economic fields, and cooperation of different interest groups).
- Partnership (Local Action Group as a local partnership involving various actors from the public, social and economic sectors).
- Innovation (local).
- Decentralisation of management and funding
- Networking and cooperation (exchange of experience and dissemination of good practices).

In its adopted assumptions, the 'Local Water Partnership' has:
1) Legal identity, e.g., a 'special' association.
2) A statute subject to public approval/consultation.

5   Directive 2006/118/EC of the European Parliament and of the Council of 12 December 2006 on the protection of groundwater against pollution and deterioration, Official journal of the European Union, 27.12.2006.
6   Council Directive 98/83/EC of 3 November 1998 on the quality of water intended for human consumption, Official journal of the European Communities, L 330/32.
7   S. Gromadzki, Model europejskiego "Lokalnego partnerstwa na rzecz wody". Realizacja celów Europejskiego Zielonego Ładu, in: M. Staniszewski, H.A. Kretka (red.), "Zrównoważony rozwój i Europejski Zielony Ład wektorami na drodze doskonalenia warsztatu naukowca", Wydawnictwo Politechniki Śląskiej, Gliwice 2021.

3)   Statutory bodies to ensure participation and equitable participation of all partners in the governance of the entity (management board, council, programme board).

4)   An office.

5)   Human resources prepared to fulfil assigned tasks,

6)   Financial and technical resources to deliver specific tasks,

7)   Leader(s) of the partnership designated by the social, public and private entities responsible for implementing, e.g., expert studies, investment tasks, participation in the crisis management system, IT, training, cooperation, maintenance of water facilities; partnership leaders (designated by the partnership) can be e.g.: local government, water company, LAG, other social and private entities.[8]

Each entity (legal entity and individual) operating within a 'Water Partnership' has the opportunity to act independently, according to its domain and vocation (assigned social, public, economic function). At the same time, by participating/participating in the "Local Water Partnership", being an integral part of it and committing its human, organisational and material resources to it, it will participate in the achievement of the overall objective and the specific objectives of its individual partners. The assumption is that this will result in the partnership – as the sum of the actions of many diverse actors – acquiring properties that none of its constituent parts have.[9] The 'Local Water Partnership' therefore replicates the functional model of the Local Action Group as a type of territorial partnership. As the preliminary studies and the experience of the European Commission show – such entities successfully implement Local Development Strategies throughout the European Union. It is recognised that LAGs can be a model for building a 'Local Water Partnership'.

Preliminary studies indicate that, for example, in the period 2004–2014 in the Podlaskie Province, the share of projects co-financed from EU funds related to water resources management was only 0.85% in the pool of all implemented projects supported by Measure 2.3 – Village renewal and preservation and protection of cultural heritage under the Sectoral Operational Programme Restructuring and modernisation of the food sector and rural development 2004–2006 and Measure Renovation and development of villages under the Rural Development Programme 2007–2013. This indicates that the rural development programmes implemented so far, despite their enormous impact on the revitalisation of many areas, both infrastructurally and socially in the countryside, have not contributed to a structural change in the water conservation, and retention system. A new approach needs to be proposed which, builds on the financial and organisational successes of the programmes

---

8   Sławomir Gromadzki, *Model europejskiego "Lokalnego partnerstwa na rzecz wody"*, 196–202.
9   Ibid.

implemented to date, as well as on projects in rural areas, to achieve the strategic objective of fulfilling the content of the European Green Deal.[10]

3   Conclusions

Based on existing solutions developed in the European Union and its Member States, it is possible to organise them relatively easily for the purposes of water conservation, and water retention. The crisis management system, which includes the conservation of water resources, flood and drought prevention, establishes the conditions for:

- water retention through the construction and expansion of reservoirs, development of irrigation and drainage systems,
- combined use of flood management structures in different areas, control of water flow in part or the whole of the state as well as water retention, and alerting of dangerous phenomena in the atmosphere and hydrosphere,
- shaping the spatial development of river valleys and floodplains, building and maintaining dykes and other hydrotechnical installations,
- the use of the Internet of Things in the aforementioned activities, providing additional opportunities to influence the environment, the economy, and the security of Poland.

A comprehensive approach to local critical infrastructure, including reservoirs and drainage and hydrological installations at the micro catchment level, incorporating them into local and regional development plans and strategies, will enable:

- The development of a comprehensive system for the management of national waters.
- The acquisition of real-time information through the use of IT tools to support information management processes, such as those relating to water retention, flood protection, detection and public alerting, and fire protection.
- Implementation of Economy 4.0 in terms of the Internet of Things (management, control, supervision, administration of critical infrastructure elements).
- Environmental protection through the development of retention in the micro catchment area to protect water resources.

---

10   Ibid.

CHAPTER 6

The Polish water retention management model within a crisis management system as a European Union Member State case study

Definition of methods and courses of action of public partners (state institutions, composite and non-complex field administrations), water companies (integrating entities of local self-government, private owners of agricultural land, forests, waters), social entities (local communities, NGOs) within the framework of state crisis management including:

– an integrated crisis management system formula covering the whole country for (macro and micro) surface water catchments,
– protecting Poland's water resources (surface and subterranean),
– the provision of water for agriculture using small and medium-sized retention in microcatchments,
– prevention of droughts, floods, contamination of watercourses, and fire protection,
– parameterisation of information from non-urbanised and urbanised areas covering micro-catchments for national security management,
– identifying practical, accessible means, tools, methods and combining them into a synergistic, coherent whole in the national security management system,
– defining organisational and normative change directions in Poland for the use of micro retention in the utilisation and protection of water resources using new techniques and technologies (e.g., situational imaging, Internet of Things) that are available and being implemented over time,

is essential for the implementation of the water retention management model in the crisis management system. The indicated elements form the basis for the creation of a universal system of water provision for agricultural production in Poland system combining, in a synergic manner, tasks in individual operational systems of the national security of the Republic of Poland: A system synergistically combining tasks within the separate operational systems of the national security of the Republic of Poland:

1. Crisis management system.
2. Flood protection system.
3. Drought management system.
4. Environmental protection system.
5. Fire protection system.

Knowledge in this area indicates directions for adapting the crisis management system at municipality, district, and provincial levels, as well as their coordination at

© JAROSŁAW GRYZ AND SŁAWOMIR GROMADZKI, 2024 | DOI:10.3920/9789004699069_008 the national level on the basis of the valorisation of hydrological data (i.e., water quality, water levels, geological data) in real time using the Internet of Things deployed in the micro-catchment environment. In addition, the identification of ways to coordinate fundamental measures, i.e., the construction of small and medium-sized retention facilities at the local level (commune and district) corresponding in size to the area of a micro catchment (in the catchment area defined for the relevant water regions), will be integrated into the crisis management system (commune, district, province, national).

This part of the work addresses questions raised in the previous chapters. It includes organisational and normative assumptions, conditions in which it may be applied, and case studies. The model was developed on the basis of the initial studies included in the book: *Przeciwdziałanie suszy, retencja wody w systemie zarządzania kryzysowego Polski.*[1]

1   Model assumptions

The proposal for a new crisis management model with a water retention subsystem is a response to the country's very adverse hydrological situation. It is the result of time-limited access to surface and subsurface water resources of a sustained nature. This state of affairs not only requires urgent action, but also continuous systemic activity throughout the country, which should be geared towards mitigating the effects of negative change. Accordingly, the approach taken is that coordination of activities should be carried out at the level of the crisis management system to ensure the appropriate standard and efficiency of the entire country.

2   Organisational assumptions

The underlying cognitive assumption is that the water retention model developed in this chapter will allow the crisis management system to function efficiently in different states of the country's operation, i.e., in normal ('usual') and emergency (crisis management) states, e.g., natural disasters, including floods, droughts, environmental contamination, and large-scale fires.

It was assumed that the construction of the model, its testing and reference to reality, its refinement on the basis of the conclusions of the research and the implementation of the operation of the water retention management system in practice would make it possible, in the long term, to achieve the following: adequate water resources, the desired level of organisation and management of water resources, the development of water storage and distribution technologies, and the utilisation

---

1   Gryz and Gromadzki S., *Przeciwdziałanie suszy*, 37.

of the resulting benefits for the development of different economic sectors. At the same time, the proposed solutions will safeguard against the effects of crisis hazards, i.e., floods and drought. As can be seen from the considerations discussed, this will allow the basic objectives of Poland's national security strategy to be achieved in the dimensions described below.

As a result of the contamination of drinking water resources, the limitation of access to water and the increase in the human population and the consequent scarcity of this resource. Accordingly:

– in the future, water will be a cause of international conflicts and threats, including but not limited to terrorism. This will result in the need for globally coordinated measures to secure access for the population to drinking water resources (as a social subsistence standard), protect water resources, build wastewater treatment plants, as well as anti-terrorist protection of water installations and infrastructure.

– access to clean water resources will become a factor in the international competitive capacity of countries in future years.[2]

The water retention measures undertaken as part of the scheme will enable the challenges outlined to be addressed. At least partially,

The proposal for a new system solution model (at central, national level) to ensure the desired level of Poland's environmental, economic and food security is a response to the dynamic nature of climate change. But the fact that they cause increasing natural disasters including droughts and floods, significant social, economic and environmental costs is not a good predictor of developments. These costs are quantifiable:

– community losses (threats to human life and health as a result of flooding, restrictions in access to drinking water as a result of drought),

– material losses (technical infrastructure, agricultural crops),

– environmental damage ("health of the environment", level of biodiversity, destruction of biocoenoses as a result of fires, such as the catastrophic fires in the Biebrza National Park in 2020, and specific hydrological changes).

– economic losses (restrictions on agricultural, industrial and energy activities, disrupted production and investment processes due to flooding or water shortages).

Ensuring sufficient and available water resources of adequate quality is the basis and prerequisite for the development of strategic sectors of the Polish economy such as agriculture, tourism, industry and energy.

Through a comprehensive, holistic approach to the problem, the model proposes organisational, legislative, financial, logistical, and technological solutions. It creates a holistic picture of the relationships and couplings to achieve the goal of

---

2 Sławomir Gromadzki, *Dostęp do zasobów wody jako jedno z uwarunkowań kategorii bezpieczeństwa*, in "Wielowymiarowość kategorii bezpieczeństwa – Ujęcie interdyscyplinarne Tom II", ed. Marcin Chełmniak, Kamil Sygidus, Piotr Kolmann (Olsztyn: Bookmarked Publishing & Editing, 2017), 192.

using the model for its intended purpose, i.e., raising the level of national security as well as development in the above-mentioned areas.

The boundaries of the research model cover the whole of Poland divided into functional and organisational areas, namely 16 provinces, 380 districts including: 314 rural districts and 66 municipalities performing the tasks of districts, so-called 'urban districts', and 2477 municipalities: 302 municipalities, of which 66 are towns with district rights, 1533 rural and 642 urban-rural[3] Taking into account 66 cities with poviat rights, the model distinguishes 2,807 independent administrative units at the voivodship, poviat and commune level. As the catchment area boundaries do not coincide with the administrative divisions of the country, the model assumes inter-municipal, inter-county, inter-provincial and cross-border cooperation to ensure that the catchment area is aligned with the boundaries of the local government units operating in their area.

In order to illustrate the model, it was assumed that the functioning subsystems in the performance of their basic tasks in relation to the crisis management system and the water retention subsystem remain autonomous. The crisis management system supports, and coordinates the operation of the water retention subsystem, causing:
- reducing the permanent state of disaster over the last five years,
- drought,
- improvement of the disaster risk management system,
- a common (nationwide/community-wide) approach to the problem,
- a systemic approach by public institutions to the issues in question.

The water retention subsystem will achieve the mission of the overarching system, i.e., the crisis management system. It will strengthen national security by introducing a new quality, first and foremost in the water management system through the "Water Partnership", which for the first time in Polish settings introduces co-responsibility of all partners, i.e., social, public and private, for the planning, financing and implementation of a nationwide water retention project. The water retention subsystem will enable the achievement of general and specific objectives – social, economic, political and strategic.

As a result of the assumptions made and the cognitive limitations in the water retention management model in the crisis management system:
- selected microcatchments,
- local authorities operating within their micro catchment boundaries,
- specific organisational and competence conditions.

By applying the model to real-world conditions, it is intended that mechanisms will be created to address its weaknesses and limitations. By refining, developing the model, organisational, functional and competence coherence can be achieved.

---

3   Liczba jednostek podziału terytorialnego kraju, Według stanu na 2020-01-01 r., accessed 03.08.2021, http://eteryt.stat.gov.pl/eteryt/raporty/WebRaportZestawienie.aspx.

The model in the crisis management subsystem within the national security system is configured in two synergistically related forms:
– flood protection.
– countering the effects of drought and complementary to them:
– counteracting environmental pollution by reducing waterborne discharges,
– preventing fires, especially large-scale fires (securing water for fire-fighting purposes),
– securing agriculture and industry with water for production,
– environmental shaping, spatial order and development, and landscape architecture.

Consequently, the proposed model is the basis for building a new and distinct system for supplying agriculture and industry with water for production.

Progression on the basis of the model takes place in two stages.

1. As part of the first (baseline) stage, an overall picture of the model of the subject under study was created, in which the most important characteristics were taken into account while eliminating features irrelevant to the subject and purpose of the research. The model developed makes it possible to explore and formulate the basic characteristics of the subject of the research and to outline a general theory about the subject under study.

2. As part of the second stage, the abstract model was brought closer to reality (through its concretisation), i.e., to the organisational, legal and operational conditions of the crisis management system of Poland and, moreover, to the realised and planned tasks in the field of water retention development within the framework of the current system of competences and standards.

The results are proposed solutions with described regularities that can be applied to the real conditions identified in the problem situation, i.e., increased risk and hazards of flooding and drought.

The application of solutions limiting the social, economic and environmental costs of eliminating the effects of natural disasters by increasing water resources will contribute to an increase in the productivity and competitiveness of the Polish economy, including in particular agriculture and the agri-food industry. These solutions will allow more effective protection of biodiversity, drinking water resources, soil and the landscape.

The model assumes that the crisis management system will be improved, thus raising the level of national security (economic, food, environmental). It is utilitarian in nature and can be used to:
– more effectively coordinate activities to increase the level of water retention in Poland,
– update, expand and improve the crisis management system,
– increase the level of national security in the areas of water management, food management and prevention of environmental risks.

The organisational model for the crisis management system, including flood protection and drought mitigation, was showcased in the implementation phase. It was divided into two stages. Stage one involves coordinating the national and local (county and municipality) crisis management system with catchment water retention planning and management. Stage two includes the identification of realistic conditions for its operation.

The model assumes coordination of the crisis management system at national (provincial) and local (county and municipality) levels. It is linked to planning and management of water retention in catchments. The assumed outcome of this action is to be:

1. To increase the effectiveness of flood risk prevention by increasing the potential of small and medium-sized retention in agricultural and forested catchments, which in turn will contribute to reducing water run-off from river and lake catchments.

2. Integration of the retention management system (planning, construction, control) into the crisis management system at state, provincial, district and municipal levels to ensure an adequate level of coordination of activities and anti-crisis planning.

3. To increase water resources in the agricultural and forestry micro catchment area in order to counteract the effects of drought and thus offset economic and environmental losses.

The model assumes:
– cooperation between the two neighbouring districts,
– cooperation between the two neighbouring provinces,
– transboundary cooperation (for this option it is necessary to diagnose and describe the water retention management system and the crisis management system in the neighbouring country, the participation of entities equivalent to Polish territorial units, as well as forms of public-private partnership, as in the case of the "Local Water Partnership")

In the area under consideration, research limitations have been assumed, i.e., only those areas of the crisis management system that relate to flood risk and drought mitigation and contamination of watercourses are considered.

3   Normative assumptions

Determining and updating the legal status of water retention in Poland's crisis management system against the background of compliance with EU and international law should include the identification of proposals for changes to existing legislation that could modify and improve existing water management at the local level with the participation of local government, water companies and other community partners. The issue is one of research and analysis of:

1. the provisions of EU law relating to water resources management, water retention management as well as the crisis management system and measures taken by governmental and self-governmental entities in this area – assessment of the state of implementation of EU law into the Polish legal system.

2. the provisions of Polish law on water resources management, water retention management and crisis management – assessment of statutory prerequisites, distinction of entities responsible for taking action, indication of legal forms and means of action, establishment of procedures for proceeding preparation of a proposal to incorporate the models developed by the research team for water retention management in the crisis management system into Polish law (in the form of legislative assumptions).

The normative assumptions for the model "Water retention subsystem in the crisis management system" are as follows:

1. Revision of the law on crisis management, which should include coordination and alignment of flood protection measures with the system for the construction of small and medium-sized retention at the national, regional (provincial) and local (district and municipality) levels.

2. The amended provisions of the law on municipal and district self-governments and the collective administration should include the imposition of new own tasks on the municipal self-government, i.e., the organisation and implementation of the tasks of the water company (as defined in the Water Law). The municipal government, in cooperation with farmers and forest owners, may actively exercise the right to initiate, create and participate in water companies as the primary form of organisation of water management in the municipality.

Furthermore, water companies operate according to the guidelines (organisational, legal and financial) set out in the Water Law. To ensure an appropriate level of local cooperation for building water retention, the municipal government participates in the creation of a "Local Water Partnership". This ensures the participation of all social, private and public partners in the planning, implementation and management of water retention. "A local water partnership", consisting of social, public and private partners, works to achieve a common social goal and the individual (particular) specific goals of the partnership members, e.g., economic, environmental, public. In view of the above, it was assumed that:

1. "The Local Water Partnership operates as a special association under the amended Associations Act.

2. For the operation of water companies, sufficient (sufficient to the desired results) funds are secured in the state budget for investment and organisational activities provided in the form of subsidies, grants, refunds.

3. It is obligatory for the municipal government to provide the legal, financial and organisational conditions for the smooth functioning of the "Local Water Partnership".

4. "The Local Water Partnership benefits from support from EU programmes, e.g., Regional Operational Programmes, Rural Development Programme, MSWiA funds, local government funds, private funds and others for the implementation of investment plans enshrined in the Water Retention Strategy and Water Retention Plan. In addition, the "Partnership" participates in administrative costs, e.g., for the functioning of the office, the employment of the necessary staff, and the handling of the day-to-day work of the "Partnership" (Board, Council).

5. The municipality, in cooperation with Polish Water (regional water management boards in the area of use of primary drainage facilities, watercourses and reservoirs owned by the State Treasury) and farmers, forest owners and other entities owning detailed drainage facilities and land/land eligible for retention construction within the framework of a formalised "Local Water Partnership", develops a municipality strategy and a water retention plan (these documents are subject to public consultation) in which it specifies:
    - the climatic and hydrological conditions of the catchment area in which the municipality is located,
    - social, private and public partners involved in the water retention management system,
    - technical, technological, environmental resources useful for building a water retention management system at municipality level,
    - strategic, strategic areas/land, protected from the point of view of building a municipal water retention management system (state and private land),
    - the principles of water resources management in the municipality, including the needs for the construction of small and medium-sized retention facilities for economic, including agricultural, purposes
    - principles for the use of primary and secondary drainage facilities and hydrotechnical devices (dams, weirs, water locks, hydroelectric power stations, canals and reservoirs, flood protection embankments) and other areas, devices, technologies and water retention techniques for controlling water run-off from catchment areas and water quality measurement devices, including for the purposes of coordinating and monitoring at municipality, district, provincial and national levels to prevent flood, ecological and water deficit risks,
    - the principles of correlation and coordination of the municipality's water retention system with the crisis management system (principles for the deployment of forces and resources) in situations of flood risk, drought risk, and complementarily water contamination and fire risk,
    - documents, including a long-term investment plan and schedule,
    - principles for the coordination of small and medium-sized water retention management activities.

The preparation of the municipal/district/provincial/national water retention strategy is based on the scientific results of atmospheric, hydrological, geological, drainage, soil morphological, technical, hydrological, phenological, dendrological and other studies, as well as expert studies, feasibility studies, and projects in individual micro-catchments in the municipalities and districts. The draft strategy developed as described above is subject to public consultation. In addition, adequate water retention strategies and plans coordinated with the crisis management system at district, provincial, and state level are being developed.

It has been accepted that the amended Crisis Management Act incorporates the 'Local Water Partnership' or their representative entities into the crisis management system. This should take place within the framework of the municipal crisis management team, possibly the municipal crisis management centre if it is functioning in the municipality in terms of flood and drought prevention or control. The water retention subsystem is identified in a formula of "Partnership" cooperation with environmental services, the State Fire Service, the Police in the area of prevention/prevention of pollution threats and elimination of the effects of contamination of watercourses, the spread of potential contaminants by limiting the outflow of water from contaminated areas and in situations of large-scale fires. The water retention subsystem included in the "Partnership" cooperation formula implements inter-municipal / district / inter-regional cooperation and is part of the coordination of transboundary measures in the case of areas where a state border runs through river basins and international cooperation is necessary to achieve the expected results.

The scope of legislative changes should include a model for water retention in the crisis management system that will enable:
– building a new financing model for the construction of water retention (financial engineering) involving the pooling of private and public funds, including EU funds and other external sources, within a common budget for all the stakeholders involved in the 'Water Partnership', i.e., social, public and private actors,
– continuous control of atmospheric and hydrological indicators down to the micro-basin level via an automated, networked and intelligent (using innovative IT solutions) real-time monitoring system, that is data acquired and processed in real time, used for ongoing decision-making in the crisis management system, and with a view to devices and vehicles configured within the framework of the Internet of Things and Economy 5.0. In addition, for the forecasting of regional, national or EU climate phenomena,
– the ongoing (real-time) management of the risk of natural disasters, e.g. floods / droughts / environmental pollution, fires, and the disposal of forces and resources, i.e. human resources seconded by the 'Partnership for Water' / deployed to cooperate / interoperate in the crisis management system and the technical infrastructure (hydro-technical, meteorological, IT) constituting the local critical infrastructure established in the micro-catchment area,

- ongoing checking, testing, improvement, of the system,
- ongoing upgrading of the skills and capabilities of the human assets involved in the system,
- ongoing maintenance (maintenance, repair, expansion) of the local critical infrastructure included in the system.

Legislative changes that recognise the implementation of the Internet of Things are essential for the success of the project, and the implementation of the model. Particularly in the aspect of:

1. Analysis of existing technical solutions in the context of their adaptability in the area indicated. The outcome here is the identification of the dedicated IoT circuits, modules and devices required, along with the associated software.
2. Implementation of an IoT module directly related to local environmental monitoring issues and retention functions.
3. Implementation of an IoT module directly related to local IoT user safety issues dedicated to small-scale retention.
4. Implementation of an IoT module directly related to crisis management for small-scale retention.

The implementation of a European model for water retention and drought management involving different forms of organisation, cooperation, coordination, responsibility, participation, public, social and private partners, institutions of the European Union and its Member States, should take into account the legislative changes indicated. These would relate to:

- water companies and their associations (operating under the terms of the Water Law),
- local authorities (municipal and district) and their associations with support and coordination from the provincial and national level,
- provincial offices (joint administration),
- provincial governments,
- regional water management boards,
- farmers and forest owners and their organisations,
- community-based organisations, e.g., Voluntary fire brigades/ firefighting units, Local Action Groups,
- entrepreneurs.

For each EU country individually, according to its specifics, and in the case of Poland with the support of:

- services, including the State Fire Service, Police, State Sanitary Inspectorate, Chief Environmental Inspectorate,
- research institutes, e.g., the Institute of Meteorology and Water Management National Research Institute, universities,
- experts,
- other stakeholders in the project, brought together in a formalised, legally owned 'Water Partnership'.

The contribution of the Water Partnership to the construction, maintenance, development of small and medium-sized water retention facilities as part of the flood and drought prevention system is linked to the concept of its pragmatic use both within the European Union and in its Member States.

4   Technical assumptions

The growth of digitalisation and automation,[4] the use of big data in spatial imaging,[5] blockchain, the implementation of Industry 4.0,[6] the internet, artificial intelligence and data mining are creating real opportunities in the area of retention control.[7] The dynamic and spatially variable nature of micro-catchments and river basins should be noted here. Control can be implemented by Real Time Control (RTC) systems to optimise retention, particularly canal retention. Implementing such systems requires local rainfall monitoring systems, local precipitation forecasts, and the metering of water levels and flows. In addition, knowledge of the projected dynamics of flow changes resulting from analysis of historical events and knowledge projections from hydrodynamic modelling work. Currently, precipitation data is being collected and archived in many places in Poland. What is lacking is their real-time processing and the application of such results to optimise the operation of hydrological systems. There is also a lack of practical application of solutions based on weather radar. Only the application of this approach can extend rainwater retention time, increase the effectiveness of rainwater management, improve the cooperation of the various entities implementing water retention, and drought prevention tasks. It should be emphasised that this type of system has been implemented in the USA for several years (Opti Platform solutions). With the introduction of 5G technology, implementing water retention, water conservation and drought prevention will become even easier. The metering of microcatchments will allow:
- Determination of variability in water availability.
- Determination of the quality of a significant body of water.
- Designing appropriate hydrological solutions.

---

4   Ian Miles, Jennifer Cassingena Harper, Luke Georghiou, Michael Keenan, Rafael Popper, The many faces of foresight, in the handbook of technology foresight. concepts and practice, ed. Luke Georghiou, Jennifer Cassingena Harper, Michael Keenan, Ian Miles, Rafael Popper (Cheltenham Massachusetts: Edward Elgar Publishing, 2008), 3–21.

5   Hong Shu, "Big data analytics: six techniques, geo-spatial information science," 19:2, (2016): 119–128, https://doi.org/10.1080/10095020.2016.1182307.

6   Heiner Lasi, Hans-Georg Kemper, Peter Fettke, Thomas Feld, Michael Hoffmann, *Industry 4.0*, Business & Information Systems Engineering, no. 4 (2014): 239–241.

7   Vincent C. Müller, Nick Boström, *Future progress in artificial intelligence: a survey of expert opinion*, in *Fundamental issues of artificial intelligence*, ed. Vincent C. Müller, Fundamental Issues of artificial intelligence (Berlin: Springer), 533–571.

– Water management due to a reduction in retention capacity (as a result of increased surface sealing, increased run-off rates and reduced wetland areas).
– Acceleration of rainwater runoff from catchment areas in urban areas.

The elements indicated will enable the use of large collections of sensors mounted on each water reservoir. In the long term, this will develop real-time control systems for micro-catchments.

5   Local water partnership

Each entity (legal entity and individual) operating within the 'Water Partnership' has the opportunity to function independently, according to its domain and vocation (assigned social, public, economic function). Thus, by participating in the 'Water Partnership', it will be an integral part of it, involving its human, organisational and material resources. This will allow the "Partnership ..." to achieve its overall objective as well as the specific objectives of the individual partners. This will result in the partnership acquiring properties that none of its constituent entities/elements have.

Using Poland as an example, it has been assumed that a "Water Partnership" is an organisational form of grouping/association/union of public, social, private partners, the aim of which is to carry out specific tasks for the benefit of:

1. Development and implementation of a municipal / district / provincial water retention strategy.
2. Development and implementation of a municipal / district / provincial water retention plan.
3. Development and implementation of a municipal / district / provincial plan for supplying agriculture with water for production purposes.
4. Coordination of water retention construction activities.
5. Coordination of activities related to the construction of an agricultural water supply system intended for production.
6. Implementation of expert tasks (studies, plans, maps, feasibility studies, projects).
7. Implementation of investment tasks.
8. Implementation of training and education activities.
9. Cooperation/participation in the crisis management system.

A Water Partnership has:

1. Legal identity such as a 'special' association.
2. Statutes subject to approval / public consultation.
3. Statutory bodies to ensure participation and equitable participation of all partners in the governance of the entity (management board, council, programme board).

4. An office.
5. Human resources prepared to fulfil assigned tasks,
6. Financial and technical resources to deliver specific tasks,
7. Leader(s) of a partnership identified by social, public and private entities responsible or accountable for the implementation for example expert studies, investment tasks, participation in the crisis management system, IT, training, cooperation, maintenance and upkeep of water facilities. The leader(s) of the partnership (appointed by the partnership) can be the local government, the water company, the LAG, other social and private entities.

6   Territorial conditions for the operation of the model

The basic functions of the presented water retention and drought prevention model will be implemented within the municipal boundaries. It is here that the basic prevention measures against the risks of natural disasters such as drought and flooding and the associated construction of small and medium-sized retention facilities take place at local level: rural commune, town, city with district rights, where the micro catchments are located (in the catchment area defined for the relevant water regions).

6.1   *The role of local government in the "local water partnership" model*
In building the water retention model, reference was made to the actual conditions in a specific Polish municipality. This is the municipality of Mały Płock, located in the Kolno district of Podlaskie Province. By referring to the real, actual and current conditions in a specific municipality, the model's assumptions, despite its idealistic approach, are fundamentally realistic, i.e., developed on the basis of the specific hydrological, administrative, environmental, social and economic conditions of a specific local authority operating in a specific catchment area.

The choice of the municipality of Maly Plock as the area on the basis of which the assumptions of the subject of the study, i.e., the model of the crisis management system with the subsystem of water retention, drought prevention, were built was driven by the results of preliminary research. Based on these, the following assessments were made:
1. It is possible to realistically apply the assumptions of the water retention, drought prevention model.
2. The municipality undertakes innovative social and economic projects, including local government cooperation in social-private-public partnerships.
3. The municipality has a great deal of experience in implementing projects subsidised by external funding, including that from the European Union.
4. Water shortages (periodic droughts) occur in the municipality, causing economic losses for the agricultural sector.

5.   The municipality is located in the catchment area of the rivers Narew, Skroda, Cetna and numerous microcatchments (mainly elements of basic and detailed drainage) feeding the watercourses indicated above.

6.   There are areas in the municipality that are favourable for the construction of small and medium-sized retention, mainly on land owned by individual farmers and forest owners.

7.   There is no system of water organisation and management in the municipality, e.g., there are no water companies, but there is an organisational, technical potential for their creation (farmers perceive that there is a need in the long term for organisational and technical solutions to ensure that their fields are irrigated). There are also remnants of drainage infrastructure built during the communist era, which could be reused after restoration, reinstatement or reconstruction.

8.   The municipality provides drinking water supply through municipal water treatment plants and waterworks (as its responsibility).

9.   The municipality is not actively involved in any form of water supply to farmers for irrigation, water retention, and is not a member of or working with any water company.

10.   The municipality, through the municipal office and the municipal utility, has the organisational, logistical and financial potential to undertake measures to build a water retention system.

11.   The municipality has the organisational capacity (experience) to build local and inter-municipal partnerships. It is also home to the headquarters of the Local Action Group "Kraina Mlekiem Płynąca" ((Land of Flowing Milk) (a partnership comprising eight municipalities): Mały Płock, Grabowo, Stawiski, Turośl, Zbójna, Nowogród, Kolno and the City of Kolno and many social and private entities) – its experience in implementing the Local Development Strategy, acting as an intermediary unit in initiating rural development measures (e.g. village renewal, diversification towards non-agricultural activities, support for micro-enterprises and others) can be used to develop the 'Water Partnership'.

12.   A Local Development Strategy is being implemented in the municipality.

13.   There is a municipal crisis management team based in the municipal office.

14.   A municipal crisis management plan is being developed.

15.   The municipality is traversed by the busy Warsaw-Mazury route, i.e., the DK-63 road, where, due to the volume of traffic, numerous accidents and road collisions occur, causing a high risk and the possibility of road traffic disasters that could lead to contamination spreading via surface and subsurface water.

16.   In the municipality there are voluntary firefighting units affiliated to the National Rescue and Firefighting System (e.g., OSP Mały Płock, OSP Kąty, OSP Rogienice, and OSP Chludnie). These units are actively involved in dealing with the consequences and risks arising from transport and natural

disasters. They are also prepared to perform new tasks, such as participating in water retention management.

17. Within the municipality, agricultural activities (crop and animal production, and services) are undertaken by farms (individual farmers) interested in participating in organised forms of water supply for agricultural production (irrigation), have land with retention potential (on which small and medium-sized retention could be built), and are interested in benefiting from the measures of the Rural Development Programme for support in:
    – farm modernisation for investment in irrigation systems,
    – investments to protect waters from pollution by nitrates from agricultural sources,
    – investments to increase the resilience of forest ecosystems and their environmental value.[8]

In the model for water retention implementation, drought prevention, within the framework of crisis management based on the potential of the Municipality of Maly Plock, it was assumed that:

1. In the municipality of Maly Plock, as a result of the social, economic, political (local politics) need to reduce the risk of the consequences of drought, as well as the need to ensure a general supply of water for agricultural production (irrigation), a "Local Water Partnership" will be established on the basis of functioning legal regulations.
2. The 'Partnership' will comprise:
    – the municipality of Maly Plock and its organisational units, i.e., Maly Plock Municipality Office, Municipal Crisis Management Team, and the Municipal Utilities Company,
    – Local Action Group "The Land of Flowing Milk",
    – Mali Plock Volunteer Fire Brigade,
    – water partnerships operating in the municipality,
    – farmers and forest owners,
    – entrepreneurs,
3. The partnership will have legal status, registered with the National Court Register, a board comprising representatives of all partners, operating in an open, transparent and public manner. The Council will elect a management board responsible for the day-to-day management of the association.
4. The partnership will have a statute defining all areas of activity, including rules for representation and membership of the association.
5. The partnership will have an office, the necessary staff (office manager, independent accounting post, and secretarial staff).

---

8 Agencja Restrukturyzacji i Modernizacji Rolnictwa, Nowości w harmonogramie PROW, accessed 8.05.2022, https://www.arimr.gov.pl/aktualnosci/artykuly/nowosci-w-harmonogramie-prow.html.

6.   The partnership will appoint representatives to undertake specific tasks:
     – the municipality of Mały Płock is represented by the mayor in the field of
       water management cooperation with the State Water Management Com-
       pany Wody Polskie, the Regional Water Management Board in Białystok,
       the Water Catchment Board in Giżycko, and the Water Supervision Author-
       ity in Kolno,
     – the municipality of Mały Plock, is represented by the mayor, in matters
       related to crisis events to cooperate with the Kolno District Office, the
       Kolno District Police Station, the Sanitary and Epidemiological Station in
       Kolno, and the Provincial Office, The Municipal Crisis Management Team
       and its head for interaction with the crisis management system,
     – water company and its president to cooperate with farmers and forest own-
       ers in water management and the use of basic and detailed drainage facili-
       ties, watercourses and reservoirs owned by the State Treasury as well as
       private owners and the community,
     – The Local Action Group and its management board is to cooperate with
       experts/institutes, research units, entities, and experts in the development
       of the Municipal Water Retention Strategy, the Municipal Water Retention
       Plan, and to cooperate with local governments that are part of the LAG
       (the LAG "Kraina Mlekiem Płynąca" is a partnership operating in the area
       of eight local governments in two counties, Koln and Łomża, and can per-
       form the relevant tasks in the "Local Water Partnerships" in the mentioned
       municipalities) in the framework of cooperation in the area of joint catch-
       ment and micro-catchment area activities,
     – The Municipal Utilities Company is to develop a municipal plan for supply-
       ing agriculture with water for production (irrigation), – The Volunteer Fire
       Brigade in Mały Płock is to cooperate with the District Headquarters of the
       State Fire Brigade in Kolno, and the District Police Headquarters in Kolno.
7.   The partnership will develop a municipal water retention and drought pre-
     vention strategy.
8.   The water retention, and drought prevention strategy will dovetail with the
     municipal land use plan and the municipal socio-economic development
     strategy.
9.   The following related (correlated) specific plans and strategies will emerge
     from the municipal development strategy:
     – the municipal water retention, and drought prevention plan,
     – a municipal crisis management plan with an inventory of local critical
       infrastructure,
     – a municipal plan for the supply of agricultural water for production
       (irrigation).

6.2   *Functional solutions for the water retention subsystem model in the crisis*
            management system at municipal, district, provincial, and state levels

The model presented for the implementation of water retention, and drought prevention in the conditions of a specific municipality allows the indication of a general scheme for real-time response to natural disaster risks. It is assumed here that the state's crisis management system with its water retention, and drought prevention subsystem will allow for real-time management of:

– local critical infrastructure to counteract the effects of floods and droughts,
– complementary actions to prevent and reduce the effects of environmental contamination and fires.

This will be possible thanks to the procedures developed and implemented as part of the national security management system, with the participation of European Union institutions within the framework of the water retention subsystem operating on the basis of municipal, district, provincial and national crisis management plans. In the adopted target solutions, crisis management plans for water retention will include procedures:

– cooperation and coordination activities, increasing the capabilities of the state security management system at all levels, i.e., provincial, district and municipal in relation to the "Local Water Partnership",
– for the deployment of forces and resources to deal with natural disasters, including floods and droughts,
– for the use (control) of local critical infrastructure to regulate the flow of water in watercourses and reservoirs in order to achieve the desired water status to reduce the risks of flooding, drought and, complementarily, the spread of water pollution, and securing water to extinguish large-scale fires, e.g. forests, crops, meadows, peat bogs, buildings, landfills and others.

The basis for the development of municipal, district and provincial management plans, taking into account the water retention subsystem, to counteract drought, will be the municipal, district and provincial strategies, as well as the nationwide and EU water retention strategy.

6.2.1   Municipality level

The municipal water retention, and drought prevention strategy according to the assumptions of the presented model in the long term is a primary planning document. Currently in Poland, there is no legal basis and no water retention strategy being developed at the local level, which means that the documents in place (strategies, plans) are of a general nature, not linked to the issue in question. There are no references to the actual conditions of individual microcatchments, no consideration of the contribution to the construction of water retention, or drought prevention or areas and infrastructure owned by farmers and forest holders. As accepted, the municipal water retention strategy should be:

- correlated with the municipal socio-economic development strategy, municipal land use plan, and municipal crisis management plan,
- consulted with the district crisis management team and the provincial crisis management centre, – and with the water authority(ies),
- adopted by the municipal council as a binding document under local law,
- an open document, subject to additions and updates.

In addition:
- take into account (on the basis of surveys and studies) the climatic conditions, including especially hydrological conditions, in the micro-catchments/catchments in which the municipality is located,
- include a description of the risks and challenges of the climate and hydrological situation, – set out the principles for the participation of social, private and public partners in the planning, construction and management of water retention within the framework of a "Local Water Partnership",
- identify the problem situation, the vision for the development of the municipality, the objectives and priorities for action linked to water resources,
- identify strategic areas of the municipality in terms of water retention capacities and drought prevention.
- identify human and technical needs and resources for water retention and drought management.
- identify the most desirable organisational, technical and technological solutions for water retention and drought prevention, taking into account geographical, climatic, hydrological, soil and other relevant conditions,
- include an inventory of local critical infrastructure for water retention, and drought and flood mitigation,
- be subject to public consultation,
- include a basic annex in the form of a municipal water retention and drought prevention plan with an agreed timetable and budget.

6.2.2 District level

The district water retention, and drought prevention strategy according to the assumptions of the presented model is a long-term planning document. This would include the drought prevention strategies (of the municipalities belonging to the district) in terms of water retention in the district area, including individual micro-catchments and catchments. This means that the district's water retention strategy should be:
- consistent with the district development strategy and the district crisis management plan,
- subject to consultation with the district crisis management team and the provincial crisis management centre,
- consulted with catchment management(s),

- an open document, subject to additions and updates.
- be subject to public consultation and adopted by the county council as a binding document of local law,
- detail (on the basis of surveys and studies) the climatic conditions, including in detail the hydrological conditions occurring in the microcatchments/catchments in which the district is located,
- include a description of the risks and challenges of the district's climate and hydrological situation, – define the principles for participation and cooperation of the district government in the planning, construction and management of water retention within the framework of local "Local Water Partnerships",
- define the principles for cooperation between the district and municipalities in the implementation of catchment/micro catchment area activities beyond the boundaries of one municipality, where coordination is needed to ensure consistency of action in the catchment area,
- indicate the most desirable organisational, technical and technological solutions for water retention, taking into account geographical, climatic, hydrological, soil conditions and other relevant factors beyond the area and domain of the municipality, but of strategic importance in terms of water retention for the district,
- include an inventory of local critical infrastructure for water retention, and drought and flood mitigation,
- include a basic annex in the form of a district water retention plan with an approved schedule and budget.

6.2.3   Provincial / regional level

District water retention, and drought prevention strategy according to the assumptions of the presented model is a long-term planning document. It should, in terms of provincial water retention, include individual microcatchments and catchment areas, be aligned with municipal and district strategies for water retention and drought prevention (municipalities and districts belonging to the province). This means that the water retention and drought prevention strategies of the individual municipalities and districts within the province should:

- be aligned with the provincial development strategy and the provincial crisis management plan,
- be aligned with the regional policy of the province concerned and the government's policy dealing with the effects of natural disasters, including the risks of floods and drought,
- be consulted with the provincial governor and the provincial crisis management team,
- be consulted with the regional water management authority(ies),
- be open documents, subject to additions and updates,
- be subject to public consultation and adopted by the Voivodship Board and then the Voivodship Assembly as a binding document of local law,

- define in detail (on the basis of studies and surveys) the climatic conditions, especially hydrological conditions, in the microcatchments/catchments where the province is located,
- include a description of the risks and challenges associated with the provincial climate and hydrological situation,
- describe the principles of cooperation with the provincial governor, the provincial crisis management centre and the national security centre regarding the participation of the provincial government in the construction of the provincial critical infrastructure and the implementation of the provincial water retention strategy,
- define the principles for participation and cooperation of the provincial government in the planning, construction and management of water retention within the framework of local "Local Water Partnerships" with the districts,
- specify the principles of cooperation between the province and the districts and municipalities in the implementation of catchment/micro catchment area activities beyond the boundaries of one district, where coordination is needed to ensure consistency of action in the catchment area,
- indicate the most desirable organisational, technical and technological solutions for water retention, taking into account geographical, climatic, hydrological, soil conditions and other relevant factors beyond the area and domain of the province (of supra-regional importance), having strategic significance for water retention in the area of the province,
- include an inventory of provincial critical infrastructure for water retention and drought and flood mitigation,
- include a basic annex in the form of a provincial water retention plan with an approved action schedule and budget.

6.2.4 State level

State policy on water retention and drought prevention should be implemented by the government with inter-ministerial cooperation between relevant ministries and government agencies on the basis of the National Water Retention Strategy. In the long term, this strategy should be a planning document for water retention on the territory of the state, covering individual catchment areas, aligned with municipal, district, provincial water retention strategies. This implies that the current approach:

- is not aligned with the country's policies and strategies for socio-economic development as well as[9] security and the national crisis management plan,

---

9   Retencja. Zatrzymaj Wodę! Program przeciwdziałania niedoborowi wody. Prognoza oddziaływania na środowisko projektu Programu przeciwdziałania niedoborowi wody, Zadanie 4.2 Opracowanie ostatecznej wersji prognozy oddziaływania na środowisko PPNW, Ministerstwo Infrastruktury, Warszawa, grudzień 2021 r.

- is not subject to inter-ministerial consultation and is not agreed by the government,
- is not subject to public consultation and is not endorsed by the Parliament by resolution of the Senate nor signed off by the President,
- does not align with municipal, district or provincial water retention strategies,
- does not describe the principles of regional and cross-border cooperation in the catchment area through which the border of a province or neighbouring country passes,
- is not an open document, subject to additions and updates,
- does not include a basic annex in the form of a national water retention and flood protection plan with a timetable and budget.

Based on the assumptions of the water retention and drought prevention model in the state security management system, a drought prevention responsibility scheme has been developed at all organisational levels. The organisational structure of the crisis management system with the water retention and flood protection subsystem includes the crisis management system with the overarching directing / leader – coordinating bodies: a) The Council of Ministers, the Prime Minister, the chairman of the RZZK; b) the minister in charge of internal affairs; c) other bodies of the minister of national defence, the minister in charge of foreign affairs, and the central offices subordinated to the President of the Council of Ministers in the following areas: water management, agriculture, environment, and maritime management. The water retention and drought prevention subsystem include overarching, directing authorities, i.e.:

- the Council of Ministers, the Minister responsible for internal affairs in consultation with the Minister for water management,
- ministers for the environment, maritime affairs, agriculture and rural development,
- provincial governors,
- chief administrative officers, and mayors,
- mayors.

In addition, cooperating entities:
- farmers and forest owners,
- water companies.
- "Water partnerships" (local government, water company, entrepreneur, farmer, forest owner, NGO, Voluntary Fire Brigade, others).

As a result, a formula will be created for the system of supplying agriculture with water for agricultural production.

6.2.5   European Union level

The long-term strategic vision of the European Union is to transform the formula of human life on Earth, leading to sustainable societies that use the achievements of science to reduce the impact of man, and communities, on the environment, the production of a carbon footprint, and the destruction of biodiversity, within a formula of green capitalism that promotes the economy and accompanying social behaviour in the education of present and future generations. To realise these objectives, the expenditure of European Union funds between 2021 and 2027 includes:

1. The Single Market, innovation and digital technologies in the Common Financial Framework €132.8 billion, in NextGenerationEU €10.6 billion, total €143.4 billion.

2. The Single Market, innovation and digital technologies in the Common Financial Framework €377.8 billion, in NextGenerationEU €721.9 billion, total €1099.7 billion.

3. Natural resources and the environment €356.4 billion in the joint financial framework, and €17.5 billion in NextGenerationEU, totalling €373.9 billion.[10]

Under the current institutional arrangements, the European Commission, with its right of legislative initiative, proposes new legislation, solutions which are examined and adopted by the European Parliament, and the Council of the European Union.[11] This state of affairs makes it imperative that water retention and drought prevention on a pan-European level find expression in the normative and organisational solutions it adopts. With the right management and executive agencies in place, it is able to orchestrate the right activities for this. In the context of water retention and drought prevention, support for the Commission's actions should be directed in the Committee of the Regions. Using the bottom-up initiative related to Local Action Groups, on which the 'Water Partnership' is modelled, the Committee of the Regions could support a pan-European approach to the implementation of a European water conservation model.

6.3 Financing of the water retention, and drought prevention subsystem

The model assumes the financing of tasks included in municipal, district, provincial water retention plans and the national plan, which will result from municipal, district, provincial water retention and drought prevention strategies and the national strategy in this area. The assumed tasks would be financed and implemented by:

1) "Water partnerships" at the municipal level,

2) Chairs of the district executive boards, district governments and district crisis management centres at district level,

3) The provincial governors and the provincial crisis management centres and the marshal offices for regional policy at the provincial level,

4) Regional water management boards at regional level,

5) Government Security Centre,

---

10 Multiannual Financial Framework (in commitments), https://ec.europa.eu/info/publications /multiannual-financial-framework-2021-2027-commitments_en, (accessed 13.07.2022.).

11 Jarosław Gryz, *Polityczne uwarunkowania integracji europejskiej z perspektywy bezpieczeństwa*, Przegląd Europejski, vol. 8 no. 2 (2021): 88–91.

6) Government, Parliament,

7) European Commission, European Parliament.

Funding for the tasks of the aforementioned entities would come from the following sources:

- European Union funding for 2021–2027 included in the Common Financial Framework and NextGenerationEU,
- the state budget for investments concerning the crisis management system, the construction of national, regional and local critical infrastructure, counteracting the effects of droughts and floods, the rescue and firefighting system, flood prevention infrastructure, including those co-financed under nationwide operational programmes and EU programmes,
- departmental funds for the implementation of state-owned infrastructure components held by, for example, the emergency management system, government institutions and agencies: The Institute of Meteorology and Water Management, Polish Waters and others within the framework of departmental programmes, including those co-financed from EU programmes,
- local (provincial) government funding for infrastructure elements owned by the provincial government, including through regional operational and EU programmes,
- local (district) government funding for infrastructure owned by the district government,
- local (municipal) government funds for financing municipal infrastructure, local critical infrastructure subsidised by operational programmes, and EU programmes,
- social funds for the construction of technical and social infrastructure belonging to non-governmental organisations, professional organisations and associations, e.g., local action groups, water companies, and Voluntary Fire Brigades, co-financed by operational programmes and other EU programmes, e.g. RDP,
- private funds for the construction of technical infrastructure, e.g., reservoirs, irrigation systems, etc., belonging to private individuals, e.g. farmers, forest owners, entrepreneurs, including those subsidised under operational programmes and EU programmes, e.g. RDP.

Water retention and drought prevention presupposes the financial co-responsibility of social, private and public partners in the implementation of tasks, maintenance of this subsystem. This will allow the following effects to be achieved: scale, synergy and overall impact of the investments made. Furthermore, to include elements of infrastructure, areas and human resources hitherto not used in natural disaster response (e.g., individuals, farmers, entrepreneurs, social organisations such as local action groups, water companies and other actors involved) in the water retention management and drought prevention system. Financial participation and co-responsibility for the operation of the system can bring tangible benefits to the stakeholders:

- private, i.e., farmers, water companies, through point preferences in applying for support under EU programmes, e.g. RDP, for investments if the supported technical infrastructure in the municipal, district, provincial strategy, water retention, drought prevention if considered strategic or constitutes an element of the local critical infrastructure,
- social, e.g., for local communities, by building technical infrastructure, landscaping, enhancing the quality of life creating and strengthening industries, including agri-tourism,
- economic, e.g., by laying the foundations for the introduction of a nationwide or EU-wide agricultural water supply system for production (irrigation).

The model discussed was also used to formulate general assumptions for the organisational structure of the water retention management and drought prevention system and in addition the division of competences for the preparation and operation of the state security management system with a water retention subsystem at municipality, district, provincial and central level, i.e. individual ministries, the entire state, as well as the European Union.

6.4   Building water retention in the rural areas of Poland

Currently, the identified forms of water retention and drought prevention possible in the existing organisational and normative solutions are:

1.   The farmer himself decides to build an irrigation system fed, for example, from a reservoir located on his own land. For this purpose, he draws up a water-legal declaration, to which, depending on the size of the reservoir, he attaches a water-legal report, obtains a water-legal permit or, alternatively, uses the water service of the catchment management (if the reservoir is located near a watercourse owned by the State). Until now, the farmer could benefit from a refund of 50% of the eligible costs under a measure for Modernisation of agricultural holdings by irrigation from the Rural Development Programme 2014–2020. And currently by extension of this programme for 2021–2022.[12] As a result, the irrigation infrastructure:
     - is planned and built on the basis of water and construction law,
     - is used exclusively for agricultural holding purposes,
     - is supplied solely from rainwater, runoff and possibly from elements of the detailed drainage system located on the farmland connected to the water storage reservoir, – is of limited coverage and impact,
     - is not functionally linked to retention built by public entities, e.g., local authorities, or regional water management authorities,
     - can be supplied by a water service from watercourses,

---

12   PROW przedłużony do 2022 r., accessed 13.07.2022, https://www.topagrar.pl/articles/prow-2014-2020/prow-przedluzony-do-2022-roku/.

- is not linked to the crisis management system, has no connection to area security, or use for commercial or other activities,
- is not linked to the general agricultural water supply system for agricultural production (irrigation),[13]
- can be subsidised as part of assistance for farm modernisation operations under the sub-measure Support for investment to agricultural holdings of the Rural Development Programme 2021–2022 regarding on-farm irrigation.

2. Retention, e.g. riverbed retention, is built on the basis of resources and infrastructure elements (such as watercourses, elements of basic and detailed land drainage, hydro technical infrastructure) belonging to the State Treasury, subject to water supervision, under the management of the catchment area of a regional water management board belonging to the Polish Waters Established infrastructure:
    - can be used for agricultural and forestry applications (through impact on adjacent land, seepage),
    - is fed by rainwater, run-off, seepage, watercourses, and detailed drainage features,
    - has a limited, single-point reach and impact,
    - is not functionally linked to retention built by private entities such as farms and forestry,
    - is not linked to the water retention and drought management system on a municipality, district or provincial scale or, more broadly, is not part of this subsystem on a national or European Union scale,
    - is not linked to the general agricultural water supply system for agricultural production (irrigation), According to S. Gromadzki, the common system of agricultural water supply for production (irrigation) is a new theoretical model of water management in agricultural space, which is not functioning in Poland and is linked organisationally and functionally to the model of water retention management system and drought prevention presented in this study.

3. Building water retention within a farm linked to the development of for example riverbed retention. For this option:
    - there are no strategic documents or plans (national or regional) linking the investments made in retention construction by private entities (individual farmers) and regional water management authorities,

---

13   The common system of water supply for agricultural production (irrigation) is a new theoretical model of water management in agricultural space, non-functional in Poland, organizationally and functionally related to the model of crisis management system with the participation of water retention subsystem, developed by S. Gromadzki; the definition and main assumptions of the model are presented in this work. The model as a whole will be presented in the next monograph focused on retention.

- there are no listed strategic documents, which means that this type of investment is rare and local in scope,
- there are no mechanisms in place to make it mandatory for the regional water management boards of Polish Waters to cooperate with farmers and forest owners on joint functionally related investments.

6.5   *Building of water retention in urbanised areas in Poland*

Water retention and drought prevention measures currently comprise a catalogue of projects for this purpose. These are as follows:

1. Let's replace concreted yards with green areas, which not only absorb water, but also help filter the air and lower its temperature. Let's persuade municipalities, neighbourhoods, cooperatives and housing communities to set up small parks (up to 5,000 sq. m) to provide small green spaces in the cities.
2. Let's argue for the planting of green walls on buildings and roofs (even bus stops) with plants that tolerate drought well. Replace lawns with flower meadows, which do a much better job of retaining moisture. When you don't want or can't give up your lawn – limit the frequency of mowing.
3. Plant squares, the edges of car parks and other small spaces with vegetation that retains rainwater from nearby areas and then gradually releases it back into the ecosystem. A rain garden can be set up in the city, in a garden or even, in a mini version, in a container on a balcony. And when we plant, for example, marsh marigold, purple loosestrife, common comfrey, sweet flag or arrowhead, your garden will not only be a retention reservoir, but also a rainwater treatment facility. Establish ponds, ponds and retention basins where possible. Importantly, permits and water consents are not needed for ponds and troughs that fill with rainwater.
4. Let's unseal concreted surfaces in cities and towns and replace them with permeable surfaces made of gravel, gravel bedding or crushed stone. They can be used to create avenues, footpaths, car parks, waiting areas or driveways. Instead of block paving and concrete, it is better to choose ecological grids and perforated slabs that allow water to soak into the ground.
5. Establish flower meadows instead of water-logged lawns. Plant trees and bushes on the grass margins and trees on field buffer strips. This will enable us to reduce water wastage by an average of 25%. Select appropriate tree and shrub species, e.g., small-leaved and broad-leaved lime trees, pedunculate and sessile oaks, white and brittle or goat willows. Also, plant blackthorn, hawthorn, wild rose, elderberry, guelder-rose or fruit trees on wasteland.[14]

Currently, there is no rainwater management in urbanised areas which creates opportunities for: a) solving water management problems caused by the progressive

---

14   *Stop suszy. Dobre praktyki, Pazem powiedzmy Stop Suszy*, Państwowe Gospodarstwo Wodne Wody Polskie, accessed 13.07.2022, www.stopsuszy.pl.

urbanisation of the landscape and identifying innovative methods to remedy them, b) economic use of rainwater, and c) supporting rainwater management decision-making in urbanised areas. This means that a range of projects spanning urbanised areas should be matched with those undertaken in rural areas. This is an added bonus for the proposed water retention, and drought prevention solutions.

6.6   *Application of water retention in accordance with model assumptions*

The analysis of the potential of the model entities, i.e. the municipality of Mały Płock and the farm located there, the water reclamation facilities managed by the Giżycko Water Catchment Board, has made it possible, not just under hypothetical, but also in real conditions, to determine how to apply the water retention and drought prevention model. This resulted in the identification of four possible options for intervention. They have been configured in relation to the potential needs of farmers as well as the risks of natural disasters occurring in the area. In the model presented here, the following assumptions were made regarding the planned model investment for water retention and drought prevention, as indicated:

1.   In the municipality of Maly Plock, the Maly Plock 'Local Water Partnership' was established on the basis of the assumptions described in the previous sections of this work.

2.   Action on water retention management and drought prevention is based on the Mały Płock municipal strategy. It was based on hydrological and climate studies of the catchments and micro-catchments and watercourses found in the municipality.

3.   Water retention and drought management measures are implemented on the basis of a private-public-community partnership involving:
     – individual farmer (agricultural land designated for the creation of a reservoir and irrigation system),
     – State Treasury represented by the catchment management of a specific regional water board,
     – the provincial governor and the combined administration (watercourses, elements of detailed and primary drainage and elements of the crisis management system),
     – municipal and district government (land and infrastructure, including local critical infrastructure),
     – community actors (technical and social infrastructure) brought together in a "Local Water Partnership", i.e., Voluntary Fire Brigade of Mały Płock, and the Local Action Group "Kraina Mlekiem Płynąca",
     – a water company (hypothetically operating).

4.   The infrastructure developed in the municipality is integrated into the water retention management system and drought prevention and results from the municipality's emergency management plan.

5.   The crisis management activity in the area of the municipality contains elements of local critical infrastructure of the district to which it belongs.

6.   The financing of the planned expenditure assumes the involvement of private, social and public funds (government, local government: municipal district and provincial), including from the EU.

7.   The infrastructure indicated in the model is integrated into the Mały Płock common municipal water supply system for agricultural production (irrigation).

8.   Farmers and other partners implementing an investment that is a component of an crisis management system with a sub-system for water retention and drought prevention, can apply for funding (reimbursement) of part of the eligible costs from EU operational and budget programmes and benefit from the point preference described in the section of the work on the LEADER method.

9.   Individual farmers and forest owners can benefit from state compensation in the form of, for example, additional area payments, de minimis aid, compensation, grants, in return for the assignment and incorporation of infrastructure built on the farm in:
      - common national water retention system,
      - crisis management system as part of local critical infrastructure,
      - common municipal water supply system for agricultural production (irrigation).

6.6.1   Option I – flood risk management

This option assumes the use of local micro catchments where a flood risk is identified in the catchment area (micro catchment) of a river or in the region(s) concerned. To prevent flooding, locally sited resources and facilities (storage reservoir, drainage facilities, and hydraulic equipment that form part of the local critical infrastructure) were used for the controlled (real-time, i.e., by controlling the outflow of water) retention of water draining from the micro-basin into the river where the flood states occur. It was assumed that in a critical state, i.e., a flood situation of a river in which a micro-basin is located, agricultural and forestry land would be flooded in order to temporarily contain the outflow of water. This is done to reduce the risk of flooding when safe conditions exist, and the containment of water run-off can be controlled. For flooding of agricultural and forestry areas, the holder/user is entitled to compensation in accordance with the bylaws and regulations for the estimation of losses and payment of compensation, which should be part of the interpretation of the operation of local water partnerships.

6.6.2   Option II – drought prevention

This option assumes that when there is a need for agricultural production water (irrigation), water is retained from the storage that constitutes the micro-basin. It is then part of the local critical infrastructure and the municipal agricultural water supply system for agricultural production (irrigation). In the event of low water levels causing a state of emergency, e.g. agricultural drought, or shortage of drinking water for the population at municipal and individual water sources, or a threat to the biodiversity of watercourses, water storage from micro catchments in the retention basins of the municipality, district or province (region) will be used. This will be done as part of ongoing water retention management. For this purpose, retention reservoirs, drainage facilities, hydraulic infrastructure that are part of the local critical infrastructure will be used. Real-time control of water run-off will allow water to be retained for agricultural production while also allowing the limiting of local use of water for agricultural production when there is a need to limit water retention and feed other areas of the province or the state with this resource by allowing drainage from the micro catchment to a watercourse/river where low water conditions exist and create a threat. In this situation, for potential reductions in access to water resources for production purposes as a result of the disposal by the disaster management system of the reservoir and infrastructure belonging to the farmer/forest owner, he/she is entitled to compensation according to the regulations and rules developed for the estimation of losses and payment of compensation, which should be part of the interpretation of the action of the local water partnerships.

6.6.3   Option III – local environmental contamination prevention

In the event of localised environmental pollution, e.g. with municipal/industrial waste water or petroleum as a result of a road traffic accident, chemical/biological substances, pesticides and other substances that may cause a risk to human health and life and a threat to the environment, a micro catchment or micro catchments will be used to prevent further contamination. In order to stop/reduce spillage and run-off of pollutants through a watercourse, the following local critical infrastructure will be used: retention reservoir, drainage facilities and hydraulic infrastructure. As part of the crisis management system (municipal level), the following procedures will be implemented (to control water run-off in real time):

- closure of sluice gates/weirs and use of other infrastructure elements to divert contaminated water to a storage reservoir,
- closure of a reservoir to retain/recapture contaminated water at its expense and that of adjacent areas,
- the prohibition of the use of contaminated water stored in the reservoir for production purposes,
- temporary exclusion of the reservoir from the water retention, drought prevention and agricultural water supply system for agricultural production (irrigation),
- inactivation/neutralisation of agents that pose a risk to human health and life or to the environment,
- restoration of the reservoir and the surrounding area,
- permitting the reuse of a previously contaminated reservoir for water retention and agricultural production/irrigation purposes.

For potential contamination/restriction of access to water resources for production purposes/reclamation needs as a result of the crisis management system's appropriation of the reservoir and infrastructure belonging to the farmer/forest owner, compensation is available in accordance with the rules of procedure drawn up, regulations for the estimation of losses and payment of compensation, which should be part of the interpretation of the local water partnerships.

6.6.4   Option IV – fire protection

In the event of large-scale fires, e.g., forests, grasslands, buildings and other structures that can cause a threat to human health and life and a threat to the environment, the water retention system will be used as part of the crisis management system. In order to limit the spread of a fire, the reservoir, drainage facilities and hydraulic infrastructure that form part of the local critical infrastructure will be used to control (in real time, i.e., by controlling the water outflow):

- the closure of sluices/weirs and the use of other infrastructure to divert water to the reservoir,
- use of the reservoir for fire-fighting purposes,
- prohibit the use of stored water for production purposes,
- temporary exclusion of the reservoir from the water retention, drought prevention and agricultural water supply system for agricultural production (irrigation).

For potential reductions in access to water resources for production purposes as a result of the crisis management system's use of the reservoir and infrastructure belonging to the farmer/forest owner, compensation will be due in accordance with the regulations and rules developed for estimating losses and paying compensation, which should be part of the local water partnerships' operating interpretation.

7   Conclusions

The considerations in this chapter show the applied use of water retention and drought prevention in the security management system. The solutions shown included two spheres, international and national. In the case of the former, the potential for solutions to address the issues of the study were identified. In the case of the second the cognitive intention of the study was detailed. The applied water retention and drought prevention model adopted for it distinguishes its variant approach. It assumes community development linked to economic, national and international security, in terms of economic, social, environmental development making use of specific forms of critical infrastructure. In doing so, it noted that:

1.   The model developed assumes a comprehensive approach to the problem of drought occurring in Poland, as well as in other European Union countries. In doing so, it demonstrates a new approach to tackling them through a

   universal, nationwide as well as EU-wide water retention and drought prevention system.

2. The developed model identifies and proposes the implementation of organisational and normative changes in water retention management and drought
   prevention in Poland and the European Union.

3. The basis of a universal water retention and drought prevention system is:
   - inclusion of water retention and drought prevention as a subsystem in the
     crisis management system of the municipality, the district, the province,
     the state and finally the European Union,
   - planning, construction and real-time management of water retention
     through the resources and organisational structure of the crisis manage-
     ment system,
   - basing water retention and drought prevention on a "Local Water
     Partnership", involving cooperation and shared responsibility for planning
     the financing and management of water retention and drought prevention
     linked to the crisis management system by public, private and community
     partners,
   - planning water retention and drought prevention at local level in which
     the basic planning and implementation document should be the munici-
     pal/district/provincial/national strategy and the water retention plan,
   - at micro catchment level, the most effective water retention and drought
     management measures should be implemented (for organisational and
     financial reasons), in conjunction with a common agricultural water sup-
     ply system for production – irrigation,
   - the model of water retention and drought prevention implies a functional
     combination of flood, contamination and fire risk prevention in the state
     security system.

4. The model assumes a balanced/rational approach to the new organisation of
   the system by using existing organisational structures to reduce the expenditure associated with system transformation – functionally/organising them
   into a coherent, more synergistically functioning system within the state.

5. The model envisages focusing the responsibility currently dispersed between
   ministries (Ministry of Agriculture and Rural Development, Ministry of
   Climate and Environment, Ministry of Interior and Administration)[15] for
   dealing with natural disasters, including the effects of drought and water
   retention, in a new security management formula with a correspondingly
   dedicated division of competences.

---

15   Water management issues were assigned to the Ministry of Maritime Affairs and Inland
     Navigation up to October 2020.

6. The implementation of the model into the practice of the state and the European Union provides the opportunity to quickly implement risk management in a drought/flood disaster situation in real time.

7. The identification and implementation of water retention and drought management in the prevention and recovery from natural disasters requires the elimination of the current lack of functional and organisational links between flood risk prevention (included in emergency plans and studies) and drought prevention through water retention at the micro-catchment level.

8. The crisis management system responds in real time to changing climatic and hydrological conditions, allowing action to be taken by the relevant entities on a nationwide scale, as well as on a micro-catchment scale, through the management of critical infrastructure (small and medium retention) in conjunction with flood protection (and the large retention infrastructure that functions within it), in countering the effects of water pollution, fire hazards, and especially large-scale fire hazards.

9. A real-time crisis management system makes it possible to manage, test, validate and improve the water retention and drought management subsystem by enabling the issuing of commands, recommendations, messages, and opinions in order to: improve the crisis management planning process at the municipal, district, provincial levels, and to make investments in the area of water retention and flood protection (creation and revision of municipal, district, provincial and national water retention plans as an element of the national crisis management plan).

10. Into the sub-system of water retention and drought prevention, flood protection, protection against contamination and fire, hitherto non-participants in the crisis management system will be included:
    - stakeholders, i.e., farmers and forest owners, and water companies (which are established and coordinated by municipal governments), and "Local Water Partnerships": local government, water company, farmer and forest owner, entrepreneur, and NGO,
    - resources (land, watercourses, reservoirs, detailed drainage elements) belonging to farmers and forest owners.

11. Water retention (planned and managed) at the micro catchment level (within the boundaries of the municipality and district) is the foundation for securing water for the economy, agriculture, the environment, contamination protection, flood protection, environmental protection and on a national scale encompassing the common agricultural water supply system.

12. The water retention and drought prevention model are open-ended, that is, it provides for continuous refinement and improvement of its assumptions after verification/checking of its suitability in real-world conditions.

13.   The model was developed in response to the inadequacies of the existing crisis management system and assumes the emergence of an efficient water retention and drought management subsystem.
14.   The model of the State and EU security management system with a water retention subsystem provides the foundation/basis for the construction of a common agricultural water supply system for agricultural production (irrigation).

The authors assume that the discussed model will be subjected to scientific, organisational and normative verification, moreover, it will be confronted with existing solutions concerning water retention and drought prevention, water management, crisis management, and the functioning of local government.

Summary

Water resources in both Poland and the European Union are limited. As a result of agricultural development, industry, the process of urbanisation, intensive recreation and tourism, there is constant pressure on the environment for its exploitation. At the same time, it is negatively reinforced by climate change. In general, measures taken in Poland and other European Union countries have focused on reducing flood risk and limiting emissions. In order to achieve the UN Sustainable Development Goals, the European Commission's European Green Deal priorities require a trans-disciplinary integration of knowledge focused on the use of properties of innovative, complementary ecosystems that define the relationship between humans and the environment. In addition, hydro-engineering tools in water management, Ecohydrological Solutions Close to Nature EH RBN (EH RBN, *Ecohydrological Nature-Based Solutions*). The formula for implementing such a holistic approach, and systemic solutions for water management, is to extend the existing mechanistic paradigm with an evolutionary-ecosystemic paradigm. This involves adopting a universal multi-dimensional objective for any water management investment – harmonising growing societal needs and aspirations with increasing the potential of each catchment area. Any investment and action in water management should improve 5 components simultaneously: W – water resources (quantity and quality); B – biodiversity (good ecological status according to Water Framework Directive); S – benefits to society (water for cities, agriculture, inland waterways, recreation); R – climate change adaptation; CE – culture and education, including analysis of public perception as a basis for evaluation and the creation of positive feedback loops between the improvement of water resources, the state of the environment, and the economy – WBSRCE. This approach is particularly relevant for Poland, for the other EU countries, as ecosystems – both terrestrial and aquatic – account for as much as 50% of recirculating, retaining and purifying of water.

The solutions described in the study include increasing surface water retention in terrestrial ecosystems, by restoring forest cover, field boundary afforestation and buffer eco-zones in agricultural areas that together create biodiversity – bio productivity and additionally, irrigation of agricultural areas and restoration of underground resources, wetlands and lakes. The above will enhance the retention of entire river basins and will be the most effective and low-cost solutions. They also contribute to the restitution of the integrity of processes along the river continuum and will positively improve water quality, bio productivity restitution and biodiversity. "The most relevant agro climatic parameters for assessing the determinants of irrigation drainage needs should be considered those that quantify the atmosphere-soil-plant system in relation to water scarcity."[1] The nature of the changes described in the study, which would happen with the contribution of water retention and drought prevention in the security management of Poland and the European Union indicates not only their potential direction, but also the conditions for success with the application of the presented interpretation.

The use of modern IT methods, the integration of hydrological solutions with Eco hydrological Solutions Close to Nature (EH NBS) will enable the shaping of the natural environment and the space in which people live such as in the 'Smart Blue-Green City. Green areas with reservoirs of retained water will contribute to the restoration of groundwater resources and the reduction of heat islands, affecting the health of residents, in addition to improving air quality, reducing CO_2 emissions and pollutants and increasing space for recreational use. Poland's water resources are defined by the amount of precipitation in river basins, its distribution over time and the potential for retention – natural and artificial, including the capacity to manage rainwater. The solution here is to adopt the approach of reducing surface sealing, implementing blue-green infrastructure, regenerating and creating small watercourses, and diverting water from rainwater drains to periodically flooded areas. The recommended method of rational water resource management, in view of its considerable temporal and spatial variability in Poland, is increasing retention, i.e., storing water when it is in excess and releasing it to users and the environment in periods of shortage. The total amount of stored water in existing retention reservoirs in Poland is about 4 billion m3, which is only 6.5% of the average annual river outflow volume. There is therefore very limited control of river runoff.[2]

It is important to recognise here the economies of scale that are associated with the implementation of activities and their effectiveness in achieving objectives. There is an urgent need for hydro-engineering solutions (sensu stricto Eco hydrological Solutions Close to Nature) to ensure water in adequate quantity and quality, available for nature, water supply and economic purposes This relates to organisational and normative solutions both in Poland and in the European Union resulting in structural, systemic, comprehensive, infrastructural, environmental, economic, social and political changes. In doing so, the concept of water retention and drought prevention indicates the need for a holistic approach involving the synergetic action of many institutions and stakeholders. First and foremost, farmers/businesses with public participation – the communities of the European Union. Warunkiem sukcesu jest przy tym skoordynowana kampania informacyjna i edukacyjna. Most importantly, it should take place at local level and be targeted at the local community and farmers/businesses. The involvement of the latter, together

1   Leszek Łabędzki, *Agroklimatyczne uwarunkowania potrzeb melioracji nawadniających,* Inżynieria Ekologiczna *Ecological engineering*, vol. 47 (May 2016): 199–204, DOI: 10.12912/23920629/62872.
2   Rządowy Program Strategiczny Hydrostrateg „Innowacje dla gospodarki wodnej i żeglugi śródlądowej", accessed 10.02.2023, https://adtg.pl/2023/08/31/projekty-z-budzetu-panstwa-2/.

with local authority institutions, is key to the success of the whole concept. This will result in the development of infrastructure as well as environmental, economic and security benefits for society as a whole and for the state, not just for Poland, but for all in the European Union.

Water retention and drought prevention included in nationwide and EU-wide system solutions will allow for the reduction of expenditures through, among other things, more effective planning, and investment efficiency. In addition, it will contribute to dissemination:

- automated methods for rapid quantification of river flows and water levels, which is crucial against the background of climate change and long periods of drought.
- techniques for comprehensive environmental water quality monitoring to reduce pollution run-off into seas and oceans.
- automated methods for rapid identification of the water quality of aquatic ecosystems.
- monitoring changes in biodiversity resulting from changes in climatic conditions.
- equipment to monitor the emergence of new types of contaminants, i.e., microplastics, viruses, bacteria, toxins, hormones, etc.
- new and rapid techniques to reduce the time taken to obtain information on the volumes of water resources and the state of the aquatic environment – both abiotic and biotic elements.
- methods for regenerating aquatic ecosystems based on actions in tune with nature.[3]

The economic benefits seem indisputable – from strengthening the resilience of the environment (biodiversity), to its attraction to tourism (enhancing tourism in specific regions), to the economy of the state and the organisation itself – from agriculture to new technologies (communications, sensors) and the implementation of economy 5.0 at the local level. Above all in the form of human interaction with artificial intelligence, protection of the environment through the use of renewable energy (sun, wind, water), and the elimination of waste.[4] The strengthening of local telecommunications networks is a prerequisite for the successful implementation of such solutions. Leveraging the Internet of Things for this would allow

---

3 Ibid.
4 It is easy to adopt an interpretation in which fields – from sowing to harvesting – are continuously monitored giving knowledge of the nature of plant growth, irrigation needs and pests The alignment of this type of element with water retention and drought prevention, generated by natural point sources of energy supply, is a matter of course. Given the nature of the scale – the use of drones and sensors for this – means that they will become as assistive as, say, a combine harvester. As the authors of the article indicate, the Internet of Things can be used in this context to monitor the environment (floodplains, sewage treatment plants), collect meteorological and climatic information and track animals. Mieczysław Ogórek, Piotr Zaskórski, *Internet rzeczy w integracji procesów zarządzania kryzysowego*, Zeszyty Naukowe Politechniki Poznańskiej, no 76 (2018): 201–213, DOI: 10.21008/j.0239-9415.2018.076.15.

the monitoring of the state of the environment – sensors (water, as well as soil and air), the collection and processing of information in local, regional, national, EU communication networks and the connection of digital and physical devices. Collectively, linked by infrastructures based on retention and drought prevention, they could serve a whole range of new applications and services based on usage criteria: always, everywhere, with everything (everything is able to present itself, provide communication and interact). This will add water retention and drought prevention to the security management system of Poland and the European Union:
- immediate provision of anticipated needs.
- continuous improvement of operational capabilities.
- convergence of information from different sources.
- personalisation for the farmer/entrepreneur and the provision of appropriately configured information for state institutions, ensuring their security.
- creating the conditions for development, from the individual farmer/entrepreneur to the state security system.

Harnessing the potential of the state and the European Union for development and safeguarding its security and prosperity creates all sorts of advantages that generate comparative advantage. Harnessing the potential of the state and the European Union for development and safeguarding its security and prosperity creates all sorts of advantages that generate comparative advantage. In addition, Poland and the other EU countries, can create an international policy instrument, such as the Food and Agriculture Organisation of the United Nations. This would create additional political capital through cooperation with other countries and organisations.

Water retention and drought prevention in the presented solutions create space for the creation of interdependencies between individuals, social groups, society and public institutions. The creation of a further element of interaction between these stakeholders will allow for a more effective implementation of the policies and development strategies of the State and the European Union in terms of social mobilisation linked to the achievement of objectives. Water retention and drought prevention also make it possible to configure a consolidated strategy for social and economic development with attention to the protection and enhancement of environmental biodiversity and the economic efficiency of agriculture.

The determinants indicated above clearly define the institutional factors determining the nature of retention in the crisis management system in Poland – the political institutions of the state, the way they are organised and function resulting from its regime and legislation. They form the basis for assessing and reviewing the effectiveness of its policies and strategies over time Therefore, it can be concluded that the area of knowledge outlined above constitutes the most important issues related to the development of the state, its security: the effectiveness of its maintenance and creation; the creation of appropriate norms serving this purpose; the legitimacy of actions taken on the basis of widespread social acceptance; the creation of conditions for the unity of those who govern and those who are governed in the pursuit of objectives to protect society, and its environment.

Water retention and drought prevention require a strategic vision of development from policy makers and their understanding of this potential and the directions supported, configured with policies and strategies. The measurability of quantitative and qualitative factors, make the presented model and Its implementation an undertaking that is not only feasible, but in a practically verifiable way, amenable to changes. The proposals contained in this study make this clear. Whether or not they will be configured within the framework of the formula for solutions described in this study is essentially a secondary matter – the point is not in the proposals put forward in one way or another, but in the pragmatic, long-term development of the countries of the European Union involving its citizens and translating them into the quality of life of all of its societies.

Bibliography

About green economy, UN environment programme, at: <https://www.unep.org/explore-topics/green-economy/about-green-economy> (accessed 6.07.2023).

Agencja Restrukturyzacji i Modernizacji Rolnictwa, Nowości w harmonogramie PROW, at: <https://www.arimr.gov.pl/aktualnosci/artykuly/nowosci-w-harmonogramie-prow.html> (accessed 8.05.2022).

Agriculture, forestry and fishery statistics 2019 edition (Luxembourg: Publications Office of the European Union, 2019), at: <https://ec.europa.eu/eurostat/documents/3217494/10317767/KS-FK-19-001-EN-N.pdf/742d3fd2-961e-68c1-47d0-11cf30b11489?t=1576657490000> (accessed 15.08.2022).

At a glance taking the EU's 'farm to fork' strategy forward, European Parliament, 2023, at: <https://www.europarl.europa.eu/RegData/etudes/ATAG/2021/690622/EPRS_ATA(2021)690622_EN.pdf> (accessed 30.04.2023).

ARiMR, PROW 2014–2020, at: <https://www.arimr.gov.pl/pomoc-unijna/prow-2014-2020.html> (accessed 5 September 2020).

ARiMR, PROW 2007–2013, at: <https://www.arimr.gov.pl/programy-2002-2013/prow-2007-2013.html> (accessed 5 September 2020).

ARiMR, SPO Rolnictwo, at: <https://www.arimr.gov.pl/programy-2002-2013/spo-rolnictwo-2004-2006.html> (accessed 5 September 2020).

ARiMR, Ryby 2014–2020, at: <https://www.arimr.gov.pl/pomoc-unijna/program-rybactwo-i-morze-2014-2020.html> (accessed 5 September 2020).

ARiMR, PO Ryby 2007–2013, at: <https://www.arimr.gov.pl/programy-2002-2013/program-operacyjny-ryby-2007-2013.html> (accessed 5 September 2020).

ARiMR, SPO Rybołówstwo i przetwórstwo ryb 2004–2006, at: <https://www.arimr.gov.pl/programy-2002-2013/spo-rybolowstwo-i-przetworstwo-ryb-2004-2006.html> (accessed 5 September 2020).

ARiMR, SAPARD, at: <https://www.arimr.gov.pl/programy-2002-2013/sapard.html> (accessed 5 September 2020).

Ayre, Margaret, Vivienne Mc Collum, Warwick Waters, Peter Samson, Anthony Curro, Ruth Nettle, Jana-Axinja Paschen, Barbara King, Nicole Reichelt, "Supporting and practising digital innovation with advisers in smart farming." *NJAS: Wageningen journal of life sciences* 90–91:1 (2019): 1–11, 10.1016/j.njas.2019.05.001.

Bach, D., Rosner, H., *Go green, live rich: 50 simple ways to save the earth (and get rich trying)*, (New York: Broadway Books, 2008).

Bachtler, John Colin Wren. "Evaluation of European Union cohesion policy: research questions and policy challenges." *Regional studies* 40, no. 2 (2006), 143–153, https://doi.org/10.1080/00343400600600454.

Balafoutis, Athanasios, Bert Beck, Sypros Fountas, Jurgen Vangeyte, Tamme Van Der Wal, Iria Soto, Manuel Gómez-Barbero, Andrew Barnes, Vera Eory. "Precision agriculture technologies positively contributing to ghg emissions mitigation, farm productivity and economics Sustainability." *Sustainability* 9(8), no. 1339 (2017): 1–21, doi.org/10.3390/su9081339.

Barnett, Jon, Stephen Dovers. *"Environmental security, sustainability and policy."* *Pacifica review: peace, security & global change* 13:2 (2001): 157–169, https://doi.org/10.1080/713604521.

Bauer, Anja, Alexander Bogner, Daniela Fuchs. "Rethinking societal engagement under the heading of responsible research and innovation: (novel) requirements and challenges." *Journal of responsible innovation* 8:3, (2021), 342–363, https://doi.org/10.1080/23299460.2021.1909812.

Berger, Thomas, Regina Birner, Nancy Mccarthy, Jose Díaz, Heidi Wittmer. "Capturing the complexity of water uses and water users within a multi-agent framework." *Water resources management* 21(2007): 129–148.

Biodiversity strategy for 2030, European Commission environment, at: <https://environment.ec.europa.eu/strategy/biodiversity-strategy-2030_en> (accessed 30.04.2023).

Britannica, at: <https://www.britannica.com/science/water-resource> (accessed 20.10.2022).

Bronson, Kelly. "Smart farming: including rights holders for responsible agricultural innovation." *Technology innovation management review* 8 (2) (2018): 7–14, http://doi.org/10.22215/timreview/1135.

Bronson, Kelly. "Looking through a responsible innovation lens at uneven engagements with digital farming." *NJAS–Wageningen journal of life sciences* 90–91 (2019), 3–5, https://doi.org/10.1016/j.njas.2019.03.001.

Bukchin, Shira, Dorit Kerret. "The role of self-control. hope and information in technology adoption by smallholder farmers – a moderation model." *Journal of rural studies* 74(4) (2020): 160–168, DOI: 10.1016/j.jrurstud.2020.01.009.

Carolan, Michael. "Digitization as politics: smart farming through the lens of weak and strong data." *Journal of rural studies* 91 (2022): 208–216, https://doi.org/10.1016/j.jrurstud.2020.10.040.

Castrignano, A., Buttafuoco, G., Khosla, R., Mouazen, A., Moshou, D., Naud, O., eds., *Agricultural Internet of Things and decision support for precision smart farming*, (Academic Press: 2020).

Cejudo, E., Navarro, F. ed. *Neo-endogenous development in European rural areas results and Lessons* (Springer International Publishing 2020).

Chełmniak, M., Sygidus, K., Kolmann, P. ed., *"Wielowymiarowość kategorii bezpieczeństwa – Ujęcie interdyscyplinarne Tom II"*, (Olsztyn: Bookmarked Publishing & Editing, 2017).

CIRCABC, CIS work programme 2022–2024.docx, at: https://circabc.europa.eu/ui/group/9ab5926d-bed4-4322-9aa7-9964bbe8312d/library/561e8b77-e75d-42d6-86a9-16405547735f/details> (accessed 6.07.2023).

Climate change and land, an IPCC special report on climate change, desertification, land degradation, sustainable land management, food security, and greenhouse gas fluxes in terrestrial ecosystems, at: <https://www.ipcc.ch/srccl/> (accessed 11.11.2020).

Climatic water balance of Poland for the period 2021-08-01–2021-09-01, Institute of Cultivation, Fertilisation and Soil Science, National Research Institute, at: <https://susza.iung.pulawy.pl/kbw/2021,14/.> (accessed 2.02.2022).

Colliander, A.E., Stone, L.F., Sundström, V., Wilms, O. & Smits, M., *Digitizing the Netherlands: how the Netherlands can drive and benefit from an accelerated digitized economy in Europe* (Boston Consulting Group: 2016).

Communication from the Commission to the European Parliament, the Council, the European Economic and Social Committee and the Committee of the Regions a blueprint to safeguard Europe's water resources, COM(2012) 673 final, Brussels, 14.11.201.

Communication from the Commission "A clean planet for all", the European Commission strategic long-term vision for a prosperous, modern, competitive and climate-neutral economy, European Commission, COM(2018) 773 final, Brussels, 28.11.2018.

Communication from the Commission to the European Parliament, the Council, the European Economic and Social Committee and the Committee of the Regions a European strategy for data, European Commission, COM(2020) 66 final, Brussels, 19.2.2020.

Communication from the Commission to the European Parliament, the Council, the European Economic and Social Committee and the Committee of the regions a new circular economy action plan for a cleaner and more competitive Europe, COM(2020) 98 final, Brussels, 11.3.2020.

Communication from the Commission to the European Parliament, the Council, the European Economic and Social Committee and the Committee of the Regions a farm to fork strategy for a fair, healthy and environmentally-friendly food system, COM/2020/381 final, Brussels, 20.5.2020.

Communication from the Commission to the European Parliament, the Council, the European Economic and Social Committee and the Committee of the Regions pathway to a healthy planet for all EU action plan: 'Towards zero pollution for air, water and soil', COM(2021) 400 final, Brussels, 12.5.2021.

Communication from the Commission to the European Parliament, the Council, the European Economic and Social Committee and the Committee of the Regions Empty Forging a climate-resilient Europe – the new EU strategy on adaptation to climate change, European Commission, COM(2021) 82 final, Brussels, 24.2.2021.

Communication from the Commission EUROPE 2020 a strategy for smart, sustainable and inclusive growth, COM(2010) 2020, 3–32, Brussels, 3.3.2010.

Commission Decision (EU) 2017/848 of 17 May 2017 laying down criteria and methodological standards on good environmental status of marine waters and specifications and standardised methods for monitoring and assessment, and repealing Decision 2010/477/EU. Report from the Commission to the European Parliament and the Council on the implementation of the Marine Strategy Framework Directive (Directive 2008/56/EC), Brussels, 25.6.2020.

Compendium of WHO and other UN guidance on health and environment, 2022 update. Geneva: World Health Organization; 2022 (WHO/HEP/ECH/EHD/22.01), at: <https://reliefweb.int/report/world/compendium-who-and-other-un-guidance-health-and-environment?gad_source=1&gclid=CjwKCAiA6byqBhAWEiwAnGCA4J_QCBlpYZOUcokwWJAmHKuvFHQXogXMEFgyAOozJ1MihJZTy2QONhoCP_YQAvD_BwE> (accessed 6.07.2023).

Consolidated version of the treaty on the functioning of the European Union, part three: union policies and internal actions – title XX: environment, Official journal 115, 09/05/2008, 0132–0133.

Council Directive of 12 December 1991 concerning the protection of waters against pollution caused by nitrates from agricultural sources (91/676/EEC).

Council Directive 98/83/EC of 3 November 1998 on the quality of water intended for human consumption, Official Journal of the European Communities, L 330/32.

Country profile factsheets, at: <https://water.europa.eu/marine/countries-and-regional-seas/country-profiles> (accessed 24.04.2023).

Country profiles on urban waste water treatment, at: <https://water.europa.eu/freshwater/countries/uwwt> (accessed 24.04.2023).

Data, maps and tools, at: <https://water.europa.eu/freshwater/data-maps-and-tools> (accessed 24.04.2023).

Dashboards on marine features under other policies, at: <https://water.europa.eu/marine/data-maps-and-tools/map-viewers-visualization-tools/dashboards-on-marine-features-under-other-policies> (accessed 24.04.2023).

Directive 2006/118/EC of the European Parliament and of the Council of 12 December 2006 on the protection of groundwater against pollution and deterioration, Official journal of the European Union, 27.12.2006.

Directive 2000/60/EC of the European Parliament and of the Council of 23 October 2000 establishing a framework for Community action in the field of water policy, Official journal of the European Communities 22.12.2000.

Directive (EU) 2020/2184 of the European Parliament and of the Council of 16 December 2020 on the quality of water intended for human consumption, Official journal of the European Union, 23.12.2020.

Decision (EU) 2022/591 of the European Parliament and of the Council of 6 April 2022 on a general union environment action programme to 2030, Official journal of the European Union, 12.4.2022.

Decree of the Minister of Infrastructure dated 15 July 2021 on the adoption of the plan for counteracting the effects of drought, Warsaw, 3 September 2021. Item 1615, 14.

Early warnings for all the UN global early warning initiative for the implementation of climate adaptation, executive action plan 2023–2027, World Meteorological Organization, at: <https://www.preventionweb.net/media/84612/download> (accessed 12.04.2023).

Ekologia.pl, at: <https://www.ekologia.pl/wiedza/slowniki/leksykon-ekologii-i-ochrony-srodowiska/deficyt-wody> (accessed 14.08.2020).

EIP Water website, at: <https://ec.europa.eu/environment/water/innovationpartnership/index_en.htm> (accessed 20.04.2023).

Elliott, Lorraine. "Human security/environmental security." *Contemporary politics*, 21:1 (2015): 11–24, https://doi.org/10.1080/13569775.2014.993905.

Ensure availability and sustainable management of water and sanitation for all, United Nations Department of Economic and Social Affairs Sustainable Development, at: <https://sdgs.un.org/goals/goal6> (accessed 5.07.2023).

EU civil protection mechanism, at: <https://civil-protection-humanitarian-aid.ec.europa .eu/what/civil-protection/eu-civil-protection-mechanism_en> (accessed 3.07.2023).

European Council meeting (20 June 2019) – conclusions, General Secretariat of the Council, EUCO 9/19.

European Council meeting (12 December 2019) – conclusions, General Secretariat of the Council, EUCO 29/19.

European Commission, Directorate-general for the environment, *2030 biodiversity strategy: removing barriers to restoring rivers*, Publications office of the European Union, 2022, at: <https://data.europa.eu/doi/10.2779/858614> (accessed 18.04.2023).

European Green Deal striving to be the first climate-neutral continent, European Commission, at: <https://commission.europa.eu/strategy-and-policy/priorities-2019-2024 /european-green-deal_en> (accessed 30.04.2023).

Eurostat, at: <https://ec.europa.eu/eurostat/cache/digpub/keyfigures/#> (accessed 14.08.2020).

Ferguson, Peter. "The green economy agenda: business as usual or transformational discourse?" Environmental politics 24 (2015): 17–37, https://doi.org/10.1080/09644016.2014 .919748.

Fielke, Simon, Robert Garrard, Emma Jakku, Alysha Fleming, Leanne Gaye Wiseman, Bruce M. Taylor. "Conceptualising the DAIS: implications of the 'Digitalisation of Agricultural Innovation Systems' on technology and policy at multiple levels." *NJAS: Wageningen journal of life sciences*, 90–91(3) (2019): 1–3, DOI: 10.1016/j.njas.2019.04.002.

Georghiou, L., Harper, J.C., Keenan, M., Miles, I., Popper, R. ed., The Handbook of technology foresight. Concepts and practice, (Cheltenham Massachusetts: Edward Elgar Publishing, 2008).

GES assessments dashboards, at: <https://water.europa.eu/marine/data-maps-and -tools/msfd-reporting-information-products/ges-assessment-dashboards> (accessed 24.04.2023).

Good Environmental Status (GES) assessments of EU marine waters by features, at: <https://water.europa.eu/marine/data-maps-and-tools/msfd-reporting-information -products/ges-assessment-dashboards/ges-marine_waters> (accessed 24.04.2023).

Global Water Security 2023 Assessment by administrator in UNU-INWEH reports, at: <https://inweh.unu.edu/global-water-security-2023-assessment/> (accessed 13.07.2023).

Global Green New Deal policy brief, United Nations environment programme, March 2009.

Granberg, L., Andersson, K., Kovách, I. ed., *Evaluating the European approach to rural development: grass-roots experiences of the LEADER programme*, (Surrey: Ashgate Publishing 2015).

Gromadzki, Sławomir, Katarzyna Glińska-Lewczuk, Marta Śliwa. "Protection of natural and cultural heritage in rural areas of the Voivodship of Podlasie in 2004–2014 supported with EU funds within the framework of the village renewal contest." Contemporary Problems of Management and Environmental Protection, no. 11 (2015): 69–80.

Gryz, J. ed., *Strategia Bezpieczeństwa Narodowego Polski*, (Warszawa: PWN Wydawnictwa Naukowe, 2013).

Gryz, J., Nowakowska – Krystman, A., Boguszewski, Ł., *Kluczowe kompetencje systemu bezpieczeństwa narodowego*, (Wydawnictwo Difin, Warszawa 2017).

Gryz, Jarosław. *"Polityczne uwarunkowania integracji europejskiej z perspektywy bezpieczeństwa."* Przegląd Europejski, vol. 8 no. 2 (2021): 88–91.

Gryz, J., Gromadzki, S., *Przeciwdziałanie suszy. Retencja wody w systemie zarządzania kryzysowego Polski* (Warszawa: Wydawnictwo Naukowe PWN, 2021).

Ibarra, Hector, Jerry Skees. "Innovation in risk transfer for natural hazards impacting agriculture." *Environmental hazards* 7:1 (2007): 62–69, https://doi.org/10.1016/j.envhaz.2007.04.008.

Innovating for sustainable growth a bioeconomy for Europe, European Commission, directorate-general for research and innovation 2012, 8–48.

Jakku, Emma, Simon Fielke, Aysha Fleming, Cara Stitzlein. "Reflecting on opportunities and challenges regarding implementation of responsible digital agri-technology innovation." *Sociologia ruralis* 62:2, (2022): 363–388, DOI: 10.1111/soru.12366.

Janc, Krzysztof, Konrad Czapiewski, Marcin Wójcik. "In the starting blocks for smart agriculture: the internet as a source of knowledge in transitional agriculture." *NJAS: Wageningen journal of life sciences*, 90–91:1 (2019): 1–12, https://doi.org/10.1016/j.njas.2019.100309.

Jeleński, Jerzy, Uwagi do założeń do programu rozwoju retencji na lata 2021–2027 z perspektywą do roku 2030 opracowanego przez Ministerstwo Gospodarki Morskiej i Żeglugi Śródlądowej, Stowarzyszenie Ab Ovo, Kraków, ul Chodkiewicza 14, at: <http://praworzeki.eko-unia.org.pl/imgturysta/files/2019-06-21%20Uwagi%20do%20Programu%20rozwoju%20retencji%202021%20-%202030.pdf> (accessed 11.08.2020).

Kasabov, E. ed., *Rural areas: obstacles, driving forces, and options for encouragement, rural cooperation in Europe*, (Basingstoke: Palgrave Macmillan. 2014).

Klerkx, Laurens, Emma Jakku, Pierre Labarthe. "A review of social science on digital agriculture, smart farming and agriculture 4.0: new contributions and a future research agenda." *NJAS: Wageningen, journal of life sciences* 90–91:1 (2019): 3–6.

Kobyliński, Andrzej. "Internet przedmiotów: szanse i zagrożenia." *Ekonomiczne problemy usług, Zeszyty Naukowe Uniwersytetu Szczecińskiego* 112 (2014): 102.

Kołacz, Małgorzata. Wojewódzki Koordynator Lokalnych Partnerstw Wodnych – KPODR w Minikowie, Lokalne Partnerstwa Wodne (LPW) – w powiatach sępoleńskim i nakielskim – Sieć Innowacji w Rolnictwie i na Obszarach Wiejskich, at: <kpodr.pl> (accessed 3.02.2022).

Kugler, R.L., *Policy analysis in national security affaires: new methods for new era* (Washington D.C.: National Defence University Press, 2006).

Kumpulainen, Kaisu. "The discursive construction of an active rural community." *Community development journal*, 52 no. 4 (2017), 611–627, DOI:10.1093/cdj/bsw009.

Kujawsko-Pomorski Ośrodek Doradztwa Rolniczego w Minikowie, at: <https://kpodr.click meeting.com/webinar-recording/fLpW30a86> (accessed 3.02.2022).

Kunming-Montreal global biodiversity framework, conference of the parties to the convention on biological diversity, CBD/COP/DEC/15/4, 19 December 2022.

Kütting, G., Lipschitz, R. ed., *Environmental governance: power and knowledge in a local–global world*, (New York: Routledge, 2009).

Labrianidis, Lois. "Fostering entrepreneurship as a means to overcome barriers to development of rural peripheral areas in Europe." *European planning studies* 14, no. 1 (January 2006): 3–8, https://doi.org/10.1080/09654310500339067.

Lasi, Heiner Hans-Georg Kemper, Peter Fettke, Thomas Feld, Michael Hoffmann. "*Industry 4.0*". *Business & information systems engineering*, no. 4 (2014): 239–241.

Law dated 19 December 2008 on public-private partnership, Journal of laws. 2009 No. 19 item 100.

Liczba jednostek podziału terytorialnego kraju, Według stanu na 2020-01-01 r., at: <http://eteryt.stat.gov.pl/eteryt/raporty/WebRaportZestawienie.aspx> (accessed 03.08.2021).

Linnerooth-Bayer, Joanne Anna Dubel, Jan Sendzimir, Stefan Hochrainer-Stigler. "Challenges for mainstreaming climate change into EU flood and drought policy: water retention measures in the Warta River Basin, Poland." *Regional Environmental Change* 15, no. 6 (2014): 1017–1023, DOI:10.1007/s10113-014-0643-7.

Lioutas, Evangelos D., Chrysanthi Charatsari. "Innovating digitally: the new texture of practices in agriculture 4.0." *Sociologia ruralis* 62:2 (2022): 250–278, DOI: 10.1111/soru.12356.

Lokalne partnerstwa ds. wody, Susza – portal Gov.pl, at: <https://www.gov.pl/web/susza/lokalne-partnerstwa-wodne> (accessed 3.02.2022).

Lokalne Partnerstwa do Spraw Wody 2021 Łódzki Ośrodek Doradztwa Rolniczego w Bratoszewicach, at: <lodr-bratoszewice.pl> (accessed 3.02.2022).

Lokalne Partnerstwo ds. Wody w powiecie koneckim – podsumowanie projektu | Świętokrzyski Ośrodek Doradztwa Rolniczego w Modliszewicach, at: <sodr.pl> (accessed 7.02.2022).

LPW, czyli Lokalne Partnerstwo Wodne – PODR w Szepietowie, at: <https://odr.pl/2020/08/03/lpw-czyli-lokalne-partnerstwo-wodne/> (accessed 3.02.2022).

Local partnership for water in your district – SIR, at: <odr.net.pl> (accessed 7.02.2022).

Local water partnership – second meeting, at: <lodr.pl> (accessed 3.02.2022).

Long-term strategy to 2050, at: <https://ec.europa.eu/clima/policies/strategies/2050_pl#tab-0-1> (accessed 10.08.2020).

Lowndes, Vivien Helen Sullivan. "Like a horse and carriage or a fish on a bicycle: how well do local partnerships and public participation go together?" *Local government studies* 30, no. 1, (2004), 51–73, DOI:10.1080/0300393042000230920.

Luke, Timothy W. "A green new deal: why green, how new, and what is the deal?" *Critical policy studies*, 3:1 (2009): 14–28, DOI: 10.1080/19460170903158065.

Łabędzki, Leszek, Agroklimatyczne uwarunkowania potrzeb melioracji nawadniających, *Inżynieria Ekologiczna Ecological Engineering*, vol. 47 (May 2016): 199–204, DOI: 10.12912/23920629/62872.

Macken-Walsh, Anne. "Partnership and subsidiarity? A case-study of farmers' participation in contemporary EU governance and rural development initiatives." *Rural society* 21, no. 1 (October 2011), 43–53, https://doi.org/10.5172/rsj.2011.21.1.43.

Maganda, Carmen. "Water security debates in 'safe' water security frameworks: moving beyond the limits of scarcity." *Globalizations*, 13:6 (2016): 683–701.

McNeill, J.R, *Something new under the sun: an environmental history of the twentieth century world*. (New York: Norton, 2000).

Milly, Paul C.D., K.A. Dunne, *Projected percentage changes in runoff, 21st century. An ensemble of 12 climate models participating in the 3rd phase of the coupled model inter-comparison project*, Geophysical fluid dynamics laboratory, Princeton University, at: <https://www.gfdl.noaa.gov/wp-content/uploads/pix/user_images/pcm/runoff_change_animation.gif> (accessed 10.08.2020).

Ministerstwo Infrastruktury, Plan przeciwdziałania skutkom suszy – Ministerstwo Infrastruktury, Portal Gov.pl at: <www.gov.pl> (accessed 4.02.2022).

Ministerstwo Gospodarki Morskiej i Żeglugi Śródlądowej, at: <https://www.gov.pl/web/gospodarkamorska/rzad-przyjal-zalozenia-do-programu-rozwoju-retencji> (accessed 11.08.2020).

MSFD reporting data explorer, at: <https://water.europa.eu/marine/data-maps-and-tools/msfd-reporting-information-products/msfd-reporting-data-explorer> (accessed 24.04.2023).

Multiannual financial framework (in commitments), at: <https://ec.europa.eu/info/publications/multiannual-financial-framework-2021-2027-commitments_en> (accessed 13.07.2022.).

Müller, V.C. ed., *Fundamental issues of artificial intelligence* (Berlin: Springer, 2016).

Nguyen, Long Le Hoang, Arlence Halibas, Trung Quang Nguyen, Determinants of precision agriculture technology adoption in developing countries: a review, *Journal of crop improvement* (2022): 11–13.

Ogórek, Mieczysław, Piotr Zaskórski. "Internet rzeczy w integracji procesów zarządzania kryzysowego." *Zeszyty Naukowe Politechniki Poznańskiej*, no 76 (2018): 201–213, DOI: 10.21008/j.0239-9415.2018.076.15.

Pauschinger, Dennis, Francisco R. Klauser, "The introduction of digital technologies into agriculture: space, materiality and the public–private interacting forms of authority and expertise," *Journal of rural studies* 91 (2022): 217–226, https://doi.org/10.1016/j.jrurstud.2021.06.015.

Polskie Radio PL, Portal Polskiego Radia, audio at: <https://www.polskieradio.pl/39/248/Artykul/633313,Powodz-tysiaclecia> (accessed 11.08.2020).

Powell, David, Lukasz Krebel, Frank van Lerven, Five ways to fund a Green New Deal. We can afford it. We can't afford not to, New economics foundation, 28 November 2019, at: <https://neweconomics.org/2019/11/five-ways-to-fund-a-green-new-deal> (accessed 30.04.2023).

Pörtner, Hans-Otto, Debra C. Roberts, Melinda Tignor, Elvira Poloczanska, Katja Mintenbeck, Anders Alegría, Marlies Craig, Stefanie Langsdorf, Sina Löschke, Vincent Möller, Andrew Okem, Bardhyl Rama (eds.), IPCC, 2022: Climate change 2022: impacts, adaptation and vulnerability. contribution of working group II to the sixth assessment report of the intergovernmental panel on climate change, (Cambridge University Press, Cambridge, UK and New York, NY, USA), 3056, doi:10.1017/9781009325844.

Proposal for a directive of the European Parliament and of the Council amending Directive 2010/75/EU of the European Parliament and of the Council of 24 November 2010 on industrial emissions (integrated pollution prevention and control) and Council Directive 1999/31/EC of 26 April 1999 on the landfill of waste, European Commission, COM(2022) 156 final/3 Strasbourg, 5.4.2022.

Proposal for a directive of the European Parliament and of the Council concerning urban wastewater treatment (recast), European Commission, COM(2022) 541 final, Brussels, 26.10.2022.

PROW przedłużony do 2022 r.,at: <https://www.topagrar.pl/articles/prow-2014-2020/prow -przedluzony-do-2022-roku/> (accessed 13.07.2022).

Rahaman, Muhammad Mizanur Olli Varis. "Integrated water resources management: evolution, prospects and future challenges." *Sustainability: science, practice and policy* 1:1 (2005): 15–21.

Regan, Áline. "Exploring the readiness of publicly funded researchers to practice responsible research and innovation in digital agriculture." *Journal of responsible innovation*, 8:1, (2021): 28–47, https://doi.org/10.1080/23299460.2021.1904755.

Regulation (EU) 2020/741 of the European Parliament and of the Council of 25 May 2020 on minimum requirements for water reuse, Official journal of the European Union, L 177/32, 5.6.2020.

Resolution adopted by the General Assembly on 25 September 2015 transforming our world: the 2030 agenda for sustainable development, General Assembly, United Nations, A/RES/70/1, 21 October 2015.

Rocznik Statystyczny Rolnictwa 2021, at: <https://stat.gov.pl/obszary-tematyczne/roczniki -statystyczne/roczniki-statystyczne/rocznik-statystyczny-rolnictwa-2021,6,15.html> (accessed 15.07.2022).

Retencja. Zatrzymaj Wodę! Program przeciwdziałania niedoborowi wody. Prognoza oddziaływania na środowisko projektu Programu przeciwdziałania niedoborowi wody, Zadanie 4.2 Opracowanie ostatecznej wersji prognozy oddziaływania na środowisko PPNW, Ministerstwo Infrastruktury, Warszawa, grudzień 2021.

Rose, David Christian, Jason Chilvers. "Agriculture 4.0: broadening responsible innovation in an era of smart farming." *Frontiers in sustainable food systems* 2 (87) (2018): 1–5, https://doi.org/10.3389/fsufs.2018.00097.

Rozporządzenie Ministra Infrastruktury z dnia 15 lipca 2021 r. w sprawie przyjęcia Planu przeciwdziałania skutkom suszy, Warszawa, dnia 3 września 2021 r. Poz. 1615.

Rozporządzenie Ministra Gospodarki Morskiej i Żeglugi Śródlądowej z dnia 11 października 2019 r. w sprawie klasyfikacji stanu ekologicznego, potencjału ekologicznego i stanu chemicznego oraz sposobu klasyfikacji stanu jednolitych części wód powierzchniowych, a także środowiskowych norm jakości dla substancji priorytetowych (Dz.U. 2019 poz. 2149).

Rozporządzenie Parlamentu Europejskiego i Rady (UE) nr 1306/2013 z dnia 17 grudnia 2013 w sprawie finansowania wspólnej polityki rolnej, zarządzania nią i monitorowania jej oraz uchylające rozporządzenia Rady (EWG) nr 352/78, (WE) nr 165/94, (WE) nr 2799/98, (WE) nr 814/2000, (WE) nr 1290/2005 i (WE) nr 485/2008, L 347/549–557.

Rozporządzenie Parlamentu Europejskiego i Rady (UE) nr 1305/2013 z dnia 17 grudnia 2013 r. w sprawie wsparcia rozwoju obszarów wiejskich przez Europejski Fundusz Rolny na rzecz Rozwoju Obszarów Wiejskich (EFRROW) i uchylające rozporządzenie Rady (WE) nr 1698/2005, L 347/487.

Rozwiązanie problemu susz i niedoboru wody w UE, at: <https://eur-lex.europa.eu/PL /legal-content/summary/addressing-water-scarcity-and-droughts-in-the-eu.html#> (accessed 5.07.2023).

Rządowy Program Strategiczny Hydrostrateg "Innowacje dla gospodarki wodnej i żeglugi śródlądowej," at: <https://adtg.pl/2023/08/31/projekty-z-budzetu-panstwa-2/> (accessed 10.02.2023).

Shu, Hong. "Big data analytics: six techniques, geo-spatial information science." 19:2, (2016): 119–128, https://doi.org/10.1080/10095020.2016.1182307.

Schikowitz, Andrea. "Creating relevant knowledge in transdisciplinary research projects – coping with inherent tensions." *Journal of responsible innovation* 7 (2) (2020): 217–234, https://doi.org/10.1080/23299460.2019.1653154.

Schilling, Janpeter, Sarah Louise Nash, Tobias Ide, Jurgen Scheffran, Rebecca Froese, Pina von Prondzinski. "*Resilience and environmental security: towards joint application in peacebuilding.*" *Global change, peace & security* 29:2 (2017): 109–114, DOI: 10.1080/14781 158.2017.1305347.

Scott, M., Gallent, N., Gkartzios, M. ed., *The routledge companion to rural planning a handbook for practice*, (Oxon/New York, NY: Routledge, 2019).

Shepherd, Mark, James A. Turner, Bruce Small, David Wheeler. "Priorities for science to overcome hurdles thwarting the full promise of the 'digital agriculture' revolution." *Journal of the science of food and agriculture* 100 (14) (2018):, 5083–5092, DOI: 10.1002/jsfa.9346.

Sisto, Roberta, Antonio Lopolito, Mathijs van Vliet. "Stakeholder participation in planning rural development strategies: using back casting to support Local Action Groups in complying with CLLD requirements." *Land Use Policy* 70 (2018), 442–450, https://doi.org/10.1016/j.landusepol.2017.11.022.

Seminarium "Przeciwdziałanie suszy" z udziałem ministra klimatu i środowiska, at: <https:// www.facebook.com/MKiSGOVPL/videos/929728354236565/UzpfSTExODc3MjU4NDgx MTczOTooMDk5NzA2NTEzMzgoOTcz/> (accessed 2.02.2022).

Sendai framework for disaster risk reduction 2015–2030, (UNISDR: Switzerland 2015), 12–37.

Situation of combined drought indicator in Europe – 2nd ten-day period of June 2023, European Drought Observatory, at: <https://edo.jrc.ec.europa.eu/edov2/php/index.php ?id=1000> (accessed 6.07.2023).

Staniszewski, M., Kretka, H.A. ed., *Zrównoważony rozwój i Europejski Zielony Ład wektorami na drodze doskonalenia warsztatu naukowca*, (Gliwice: Wydawnictwo Politechniki Śląskiej, 2021).

Steinke, Jonathan, Jacob van Etten, Anna Müller, Beta Ortiz-Crespo, Jeske van de Gavel, Silvia Silvestri, Jan Priebe. "Tapping the full potential of the digital revolution for agricultural extension: an emerging innovation agenda." *International Journal of Agricultural Sustainability* 19:5–6, March (2020): 549–565, DOI: 10.1080/14735903.2020.1738754.

Stop suszy. Dobre praktyki, Pazem powiedzmy Stop Suszy, Państwowe Gospodarstwo Wodne Wody Polskie, at: <www.stopsuszy.pl> (accessed 13.07.2022).

Strategic foresight analysis 2017 Report, (Allied commander transformation NATO: 2017).

Struzik, Renata, Tworzone są Lokalne Partnerstwa ds. Wody, PGW Wody Polskie, at: <https://www.terazsrodowisko.pl/aktualnosci/rolnictwo-niedobory-susza-Lokalne-Part nerstwa-ds-Wody-9093.html> (accessed 3.02.2022).

Study on European Union (EU) integrated policy assessment for the freshwater and marine environment, on the economic benefits of EU water policy and on the costs of its non-implementation" (BLUE2) commissioned by the European Commission (EC) (Luxembourg: Publications Office of the European Union, 2019).

Sustainable development: towards the achievement of sustainable development: implementation of the 2030 Agenda for Sustainable Development, including through sustainable consumption and production, building on Agenda 21 Resolution adopted by the General Assembly on 21 December 2020 [on the report of the Second Committee (A/75/457/Add.1, para. 14)] 75/212. United Nations Conference on the midterm comprehensive review of the implementation of the objectives of the international decade for action, "Water for sustainable development", 2018–2028.

Sutowski, M. ed., *Ekonomia polityczna „dobrej zmiany"*, (Warszawa: Instytut Studiów Zaawansowanych: 2017).

System Monitoringu Suszy Rolniczej, Ministerstwo Rolnictwa i Rozwoju Wsi, at: <https://susza.iung.pulawy.pl/> (accessed 21.12.2020).

Szymańska, U., Falej, M., Majer, P. ed., *Organizacje pozarządowe w ujęciu prawno-postulatywnym*, (Olsztyn: Wydział Prawa i Administracji UWM, 2017).

Taherzadeh, Mohammed J. "Bioengineering to tackle environmental challenges, climate changes and resource recovery." *Bioengineered*, 10:1 (2019): 698–699.

The European Green Deal, communication from the Commission to the European Parliament, the European Council, the Council, the European Economic and Social Committee and the Committee of the Regions, Brussels, 11.12.2019 COM(2019) 640 final.

The Municipality of Maly Płock and its location within Poland, Wikipedia, at: <https://pl.m.wikipedia.org/wiki/Plik:Ma%C5%82y_P%C5%82ock_(gmina)_location_map.png> (accessed 5.10.2022).

The United Nations world water development report 2019: leaving no one behind, United Nations Educational, Scientific and Cultural Organization, Paris (2019): 1–34.

Tworzenie Lokalnych Partnerstw ds. Wody (LPW) w Województwie Opolskim | Opolski Ośrodek Doradztwa Rolniczego, at: <oodr.pl> (accessed 3.02.2022).

Tworzone są Lokalne Partnerstwa ds. Wody PGW Wody Polski, at: <teraz-srodowisko.pl> (accessed 7.02.2022).

Uncovering pathways towards an inclusive Green Economy A Summary for Leaders, United Nations Environment Programme, 2015, at: <https://wedocs.unep.org/bitstream /handle/20.500.11822/9838/-_Uncovering_Pathways_towards_an_Inclusive_Green _Economy_a_Summary_for_Leaders-2015IGE_NARRATIVE_SUMMARY_Web.pdf.pdf

?sequence=3&amp%3BisAllowed=y%2C%20Portuguese%7C%7Chttps%3A//wedocs .unep.org/bitstream/handl> (accessed 6.07.2023).

United Nations Secretary-General's plan: water action decade 2018–2028, at: <https://sdgs .un.org/sites/default/files/2021-05/UN-SG-Action-Plan_Water-Action-Decade-web_0 .pdf> (accessed 7.04.2023).

2023 United Nations conference on the midterm comprehensive review of the implementation of the objectives of the international decade for action, "Water for sustainable development", 2018–2028, United Nations, A/CONF.240/2023/8, 31 January 2023.

UNESCO, UN-Water, 2020: United Nations world water development report 2020: water and climate change, (Paris: UNESCO).

Ustawa z dnia 20 lipca 2017 r. prawo wodne, Dz.U.2020.0.310.

Valsson, T., *How the world will change global warming*, (Reykjavík: University of Iceland Press, 2006).

Warmia and Mazury Agricultural Advisory Centre in Olsztyn, at: <wmodr.pl> (accessed 7.02.2022).

What is water security? Infographic, at: <https://www.unwater.org/publications/what-water -security-infographic> (accessed 14.07.2023).

Water notes – about integrated water management, EU water legislation and the Water Framework Directive, at: <https://ec.europa.eu/environment/water/participation/notes _en.htm> (accessed 5 September 2020).

Water resource management – Local Water Partnerships (LPW), at: <https://modr.pl /aktualnosc/zarzadzanie-zasobami-wody-lokalne-partnerstwa-ds-wody-lpw> (accessed 3.03.2022).

Water note 1 joining forces for Europe's shared waters: coordination in international river basin districts, European Commission (DG Environment), March 2008, at: <https:// ec.europa.eu/environment/water/participation/pdf/waternotes/water_note1_joining _forces.pdf> (accessed 18 April 2023).

Water note 2 cleaning up Europe's waters: identifying and assessing surface water bodies at risk, European Commission (DG Environment) March 2008, at: <https://ec.europa .eu/environment/water/participation/pdf/waternotes/water_note2_cleaning_up.pdf> accessed (accessed 18 April 2023).

Water note 3 groundwater at risk. managing the water under us, European Commission (DG Environment) March 2008, at: https://ec.europa.eu/environment/water/participa tion/pdf/waternotes/water_note3_groundwateratrisk.pdf> (accessed 18 April 2023).

Water note 4 reservoirs, canals and ports: managing artificial and heavily modified water bodies, European Commission (DG Environment) March 2008, at: <https://ec.europa .eu/environment/water/participation/pdf/waternotes/water_note4_reservoirs.pdf> (accessed 18 April 2023).

Water note 5 economics in water policy: the value of Europe's waters, Commission (DG Environment) March 2008 at: <https://ec.europa.eu/environment/water/participation /pdf/waternotes/water_note5_economics.pdf> (accessed 18 April 2023).

Water note 6 monitoring programmes: taking the pulse of Europe's waters, Commission (DG Environment) March 2008 at: <https://ec.europa.eu/environment/water/partici pation/pdf/waternotes/water_note6_monitoring_programmes.pd> (accessed 18 April 2023).

Water note 7 intercalibration: a common scale for Europe's waters, Commission (DG Environment) March 2008, at: <https://ec.europa.eu/environment/water/participa tion/pdf/waternotes/water_note7_intercalibration.pdf> (accessed 19 April 2023).

Water note 8 pollution: reducing dangerous chemicals in Europe's waters, Commission (DG Environment) December 2008, at: <https://ec.europa.eu/environment/water/participa tion/pdf/waternotes/water_note8_chemical_pollution.pdf> (accessed 20 April 2023).

Water note 9 water notes on the implementation of the Water Framework Directive integrating water policy: linking all EU water legislation within a single framework, Commission (DG Environment), December 2008, at: <https://ec.europa.eu/environ ment/water/participation/pdf/waternotes/water_note9_other_water_legislation.pdf> (accessed 20 April 2023).

Water note 10 water notes on the implementation of the Water Framework Directive climate change: addressing floods, droughts and changing aquatic ecosystems, Commission (DG Environment) December 2008, at: <https://ec.europa.eu/environment/water /participation/pdf/waternotes/water_note10_climate_change_floods_droughts.pdf> (accessed 20 April 2023).

Water note 11 water notes on the implementation of the Water Framework Directive from the rivers to the sea: linking with the new Marine Strategy Framework Directive, Commission (DG Environment) December 2008, at: <https://ec.europa.eu/environ ment/water/participation/pdf/waternotes/water_note11_marine_strategy.pdf> accessed (accessed 20 April 2023).

Water note 12 a common task: public participation in river basin management planning, Commission (DG Environment) December 2008, at: <https://ec.europa.eu/environment /water/participation/pdf/waternotes/water_note12_public_participation_plans.pdf> (accessed 20.04.2023).

Water resource management – Local Water Partnerships (LPW), at: <https://modr.pl /aktualnosc/zarzadzanie-zasobami-wody-lokalne-partnerstwa-ds-wody-lpw> (accessed 3.03.2022).

Water, sanitation and hygiene (WASH), at: <https://www.who.int/health-topics/water -sanitation-and-hygiene-wash#tab=tab_1> (accessed 5.07.2023).

Water scarcity and droughts. Preventing and mitigating water scarcity and droughts in the EU, at: <https://environment.ec.europa.eu/topics/water/water-scarcity-and-droughts _en> (accessed 6.07.2023).

Water scarcity prevention programme, at: <https://www.gov.pl/web/infrastruktura/pro gram-przeciwdzialania-niedoborowi-wody> (accessed 2.02.2022).

Waters, Colin N., Jan Zalasiewicz, Colin Summerhayes, Anthony D. Barnosky, Clement Poirier, Agnieszka Gałuszka, Alejandro Cearreta, Matt Edgeworth, Erle C. Ellis, Michael

Ellis, Catherine Jeandel, Reinhold Leinfelder, J.R McNeill, Daniel deB Richter, Will Steffen, James Syvitski, Davor Vidas, Michael Wagreich, Mark Williams, An Zhisheng, Jacques Grinevald, Eric Odada, Naomi Oreskes, Alexander P. Wolfe, "The Anthropocene is functionally and stratigraphically distinct from the Holocene." *Science* (2016), 351. Doi: 10.1126/science.aad2622.

Watkins, Josh. "Spatial imaginaries research in geography: synergies, tensions, and new directions." *Geography compass* 9, no. 9 (2015): 508–522, https://doi.org/10.1111/gec3.12228.

WISE freshwater resource catalogue, at: <https://water.europa.eu/freshwater/data-maps-and-tools/metadata> (accessed 24.04.2023).

Wiingaard Stoustrup, Sune. "The re-coding of rural development rationality: tracing EU governmentality and Europeanisation at the local level." European planning studies (2021): 1–14, https://doi.org/10.1080/09654313.2021.2009776.

WMO provisional state of the global climate 2022, World Meteorological Organization, at: <https://library.wmo.int/doc_num.php?explnum_id=11359> (accessed 12.04.2023).

World Bank. 2021. "Water in agriculture." World Bank, at: <https://www.worldbank.org/en/topic/water-in-agriculture#1> (accessed 20.03.2022).

W Senacie o Lokalnych Partnerstwach ds. Wody, Dziennik Warto Wiedzieć, at: <warto wiedziec.pl> (accessed 8.02.2022).

Wsparcie dla tworzenia Lokalnych Partnerstw ds. Wody, at: <https://www.cdr.gov.pl/aktualnosci-instytucje/3367-wsparcie-dla-tworzenia-lokalnych-partnerstw-ds-wody-lpw> (accessed 3.02.2022).

von Weizacker, E.U., Lovins, A.B., Lovins, L.H., *Factor four: doubling wealth – halving resource use: a report to the Club of Rome*, (New York: Kogan Page 2001).

Zarządzanie zasobami wody – Lokalne Partnerstwa ds. Wody (LPW), at: <https://modr.pl/aktualnosc/zarzadzanie-zasobami-wody-lokalne-partnerstwa-ds-wody-lpw> (accesed 20.09.2022).

Zero pollution action plan: a first step towards a blue deal for a water-smart society, Water Europe Technology & Innovation, at: <https://watereurope.eu/wp-content/uploads/Position-paper-Zero-Pollution-Strategy-FINAL-1.pdf> (accessed 30.04.2023).

Zeweld, Woldegebrial, Guido Van Huylenbroeck, Girmay, Tesfay, Stijn Speelma. "Smallholder farmers' behavioural intentions towards sustainable agricultural practices." *Journal of environmental management* 187 (2017): 71–81, DOI: 10.1016/j.jenvman.2016.11.014.

Index